SCIENCE EXPLAINED

Colorado Academy

CHIP NEWCOM

EIGHTH GRADE CONTINUATION

June 3, 1998

SCIENCE EXPLAINED

The World of Science in Everyday Life

Colin A. Ronan, General Editor

A HENRY HOLT REFERENCE BOOK

HENRY HOLT AND COMPANY

NEW YORK

Contents

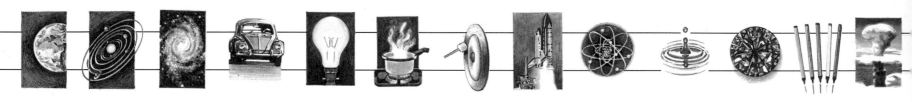

Henry Holt and Company, Inc.
Publishers since 1866
115 West 18th Street
New York, New York 10011

Henry Holt ® is a
registered trademark of
Henry Holt and Company, Inc.

**Library of Congress
Cataloging-in-Publication Data**
Science explained: the world of science
in everyday life / general editor,
Colin A. Ronan. — 1st ed.
p.   cm.   — (Henry Holt reference book)
1. Science—Miscellanea—Popular works.
I. Ronan, Colin A.
II. Series.
Q173.S42   1993   93-15439
500—dc20   CIP
ISBN 0-8050-2551-0
ISBN 0-8050-4236-9 (An Owl book: pbk.)

First published in 1993 by Henry Holt
Reference Books.

First Owl Book Edition—1996

Conceived, edited and designed by
Marshall Editions Ltd., London
Printed and bound in Italy by
New Interlitho SpA, Milan
Origination by Scantrans Pte, Singapore
Filmsetting supplied by
Dorchester Typesetting Group Limited

All first editions are printed on acid-free paper.∞

10 9 8 7 6 5 4 3 2 1
10 9 8 7 6 5 4 3 2 1 (pbk.)

Project editor   Jon Kirkwood
Art editor   Marnie Searchwell
Assistant editor   Heather Magrill
DTP editor   Mary Pickles
Copy editors   Lindsay McTeague
Maggie McCormick
Picture editor   Zilda Tandy
Picture research   Judy Lehane
Elizabeth Loving
Researchers   Helen Burridge
Jon Richards
Contributors   David Burnie
Chris Cooper
John Farndon
Steven Parker
Robin Scagell
Consultant   Dr. Phil Whitfield
Editorial director   Ruth Binney

Production   Barry Baker
Janice Storr
Nikki Ingram

Previous page clockwise from top
lifting a balloon with hot gases: atoms set
free; energy's endless flow in a car trip;
looking inside the brain: head office;
a plant with a strange sense – touch.

Overleaf clockwise from top
rocky neighbors: sister Earth and brother
Mars; modeling reality in a virtual
world; power in your pocket from a battery
generating electricity; the photon and the
flower – color from the atomic paintbox;
plants gathering light for life.

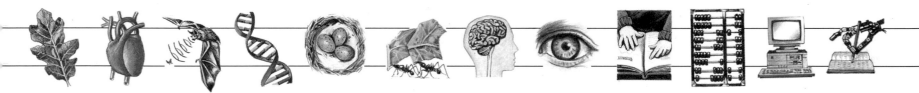

Foreword

Science plays a vital part in our lives. It meets us at every turn – at home, outdoors, while we eat, and even while we sleep.

Here we set out both to explain science, and also to broaden our understanding of it. We show how the different sciences – astronomy, chemistry, physics, and biology – are not separate, but dovetail together. From astronomy, we discover atoms in space which made our planet Earth and all it contains. By studying those atoms, we see how they behave, the forces they release, and the nature of the energy they supply. We also examine how atoms group together into molecules to make all the substances we see around us. We even find that these molecules are the stuff of life. Then putting all this knowledge together with science's latest tool – the computer – we can even discover the working of that pinnacle of evolution – the human brain.

Yet the computer does more than help science and industry. It affects our arts and leisure, thus making it a vital part of human culture. So, too, is the whole of science. Indeed, there is no field of human endeavor that is not linked to science in one way or another. The Big Bang theory of the beginning of the universe, the strange quantum world of atoms, and the discovery of genes and DNA are just three aspects of the most vivid scientific imaginations being put to work to explain the amazing results of experiment and observation. They show the wonder of the natural world, just as the painter, the poet, the musician, and others in the world of the arts speak to us of the wonders which they see. Our true culture must neglect none, but embrace them all.

Colin A. Ronan

Introduction

The pressure to divide and subdivide science into isolated specialist subjects has created a situation in which the broad vision of science risks being lost. By reuniting the body of knowledge within just five interlinked sections, this book allows the full breadth of understanding to be restored. What is more, the topics brought into focus in each section do not confine the enquiring mind to narrow fields of thought. Instead, they explain important principles in their true context – the context of everyday reality.

SPACE
The night sky is not merely an awe-inspiring sight, it also holds the key to our past – and our future. From life on Earth to the life cycle of the universe, **Space** looks at the science of our planet, the solar system, stars, and galaxies and explores the exotic objects scattered throughout the heavens.

ENERGY
Never created and never destroyed, **Energy** is the most convertible currency in the universe. Flowing from one diverse form to another, it powers the many processes of existence. Understanding its many aspects and the way it is used in machines and by living things is central to science.

ATOMS AND MATTER
At the fundamental level, our knowledge of the world about us comes from the science of **Atoms and Matter.** This section shows how atoms are constructed, how they work, and how they link up in chemical reactions. It also deals with the manifestations of the forces that bind matter together.

LIFE
The huge variety of living creatures survives by harvesting energy from the environment. **Life** explores the diversity of the living world and delves into its most intimate corners. A grasp of life processes allows a full appreciation of the richness of nature.

BRAINS AND COMPUTERS
From wetware to hardware to software, **Brains and Computers** steadily uncovers the layers of complexity at the frontiers of thought – both biological and electronic. Encompassing perception, memory, and consciousness, it also probes deep into the silicon heart of the computer.

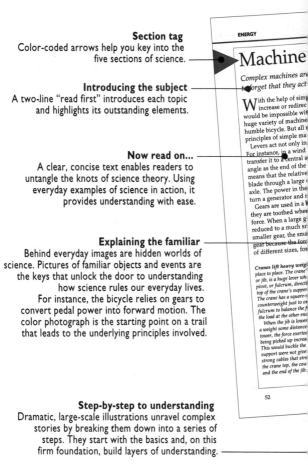

Section tag
Color-coded arrows help you key into the five sections of science.

Introducing the subject
A two-line "read first" introduces each topic and highlights its outstanding elements.

Now read on...
A clear, concise text enables readers to untangle the knots of science theory. Using everyday examples of science in action, it provides understanding with ease.

Explaining the familiar
Behind everyday images are hidden worlds of science. Pictures of familiar objects and events are the keys that unlock the door to understanding how science rules our everyday lives. For instance, the bicycle relies on gears to convert pedal power into forward motion. The color photograph is the starting point on a trail that leads to the underlying principles involved.

Step-by-step to understanding
Dramatic, large-scale illustrations unravel complex stories by breaking them down into a series of steps. They start with the basics and, on this firm foundation, build layers of understanding.

HOW TO USE THIS BOOK
Photographs and illustrations of everyday things are the starting point that leads to the principles of science. Words link carefully to pictures to explain in depth the science of our world.

Each of the book's double-page spreads is a self-contained story. But the world of science is complex, and topics do not always fit neatly and completely under the headings superimposed on them. To deal with this, connections lead from the edge of each right-hand page to other topics, both within the same section and between different sections.

In order to promote flexibility of thought and depth of interest, the connections made are often deliberately wide-ranging. Using the connections will make the book fully interactive and forge the links of understanding between the different branches of science.

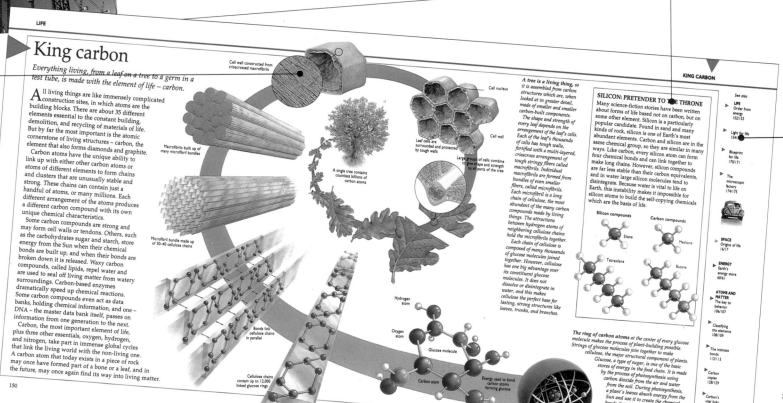

Getting the picture
Large-scale illustrations investigate the science of the everyday using clear, labeled images backed up with concise captions.

Connection icon
Graphic icons help you make the link between a specific topic and connecting topics in the same or different sections.

Close-up on science
Meticulous artworks explain the all-important details of science or project the topic to a fuller, more complete level.
By exploring how gears work on a bicycle, the idea of mechanical forces and how they are used is dealt with in a friendly, approachable manner.

Feature box
To expand understanding, box features show the detail of how science works, or give contrasting or complementary examples. Here, carbon is compared with its chemical cousin silicon.

Connections
Follow the routes to suggested topics that contain back-up facts to boost your grasp of each subject.
The connections also track down related themes and explore parallel pathways of knowledge, so that science becomes integrated into a coherent body of knowledge.
Linked topics in the same section are listed first, followed by topics in other sections. Each topic title is section-tagged and followed by its page number for easy access.

Space

Our vantage point in the universe is a small planet orbiting an average star in a backwater of a large galaxy that is just one of countless others scattered throughout the universe. From our home world – the only place where we know for sure that life exists – we look into space and see the wonders of the cosmos.

Close to Earth are the planets and other bodies of the solar system orbiting our familiar, life-giving Sun. Farther out are the other stars of our galaxy, some bright and hot, others tiny and dim. We can see gas clouds from which stars spring into existence and detect strange phenomena that show where stars have died in cataclysmic violence to leave holes of nothingness behind. There are milky pools that show where other galaxies exist and, stretching the tools of astronomy to their limits, scientists can probe the ultimate mysteries: how the universe could have begun – and how it may meet its end.

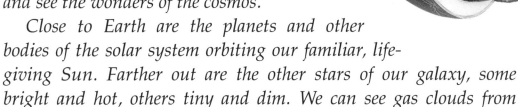

Left clockwise from top: black hole formation; our galaxy; the planets of the solar system; a stage in the evolution of life on Earth.
This page (top): *the planet Mercury;* **(left)** *global weather systems.*

Our place in space

The movements of the Sun in the sky are governed by the way the Earth moves around the Sun.

The evidence of our eyes suggests that the Sun, because it rises in the east and sets in the west, moves around the Earth. But we have known for a long time that the Earth, our home, is a basketball-shaped planet which turns on its axis once a day and moves around the Sun once a year.

At the equator the Earth's rate of spin is about 1,000 mph (1,600 km/h), and the speed of Earth's orbit around the Sun is 66,500 mph (107,000 km/h).

At night, the steady turning of the Earth is evident as the stars, like the Sun and Moon, rise and set. Earth's year-long orbit around the Sun is shown, too, in the way different stars are visible from season to season. It is as if we were on a merry-go-round: our view changes as the Earth turns.

But unlike a simple merry-go-round, Earth has an axis which is tilted in relation to the line of its orbit. This means that in some places the Sun is sometimes high in the sky at midday and sometimes low. Our seasons are a constant reminder of this tilt as the Sun travels higher in the sky in summer and lower in winter.

Earth is 7,929 miles (12,760 km) across, yet the zone in which we live – the zone of comfortable human habitation – is less than 3 miles (5 km) thick. If the Earth was an egg and this zone – from sea level up to high mountains – was its shell, the egg would be over 5¾ feet (1.75 m) across.

The Sun rises almost vertically at the equator, which means that day dawns much more quickly than in middle latitudes. At the equator the Sun passes overhead at midday in March and September, but moves north or south of this point in June or December. The seasons at the equator are much less extreme than at areas closer to the poles.

Though the Sun travels almost overhead at the equator, at mid-latitudes it rises at an angle and travels less high in the sky. A polar observer sees it rise above the horizon at the spring equinox and sink below the horizon after the autumn equinox. In fact, at the poles the Sun travels parallel to the horizon, moving only slowly north or south as the year progresses. These diagrams (below) show the northern hemisphere view. In the southern hemisphere, the Sun still rises in the east and sets in the west, but it moves across the north of the sky instead of the south.

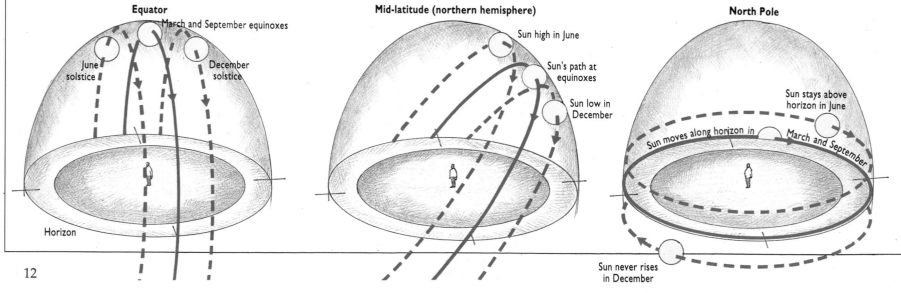

Equator
March and September equinoxes
June solstice
December solstice
Horizon

Mid-latitude (northern hemisphere)
Sun high in June
Sun's path at equinoxes
Sun low in December

North Pole
Sun stays above horizon in June
Sun moves along horizon in March and September
Sun never rises in December

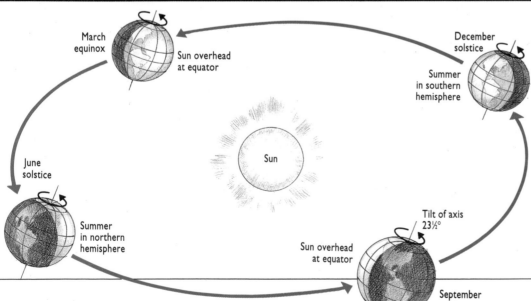

March equinox

Sun overhead at equator

December solstice

Summer in southern hemisphere

June solstice

Sun

Summer in northern hemisphere

Tilt of axis 23½°

Sun overhead at equator

September equinox

Seasons occur because Earth's axis is not at right angles to its orbit (left). In summer the Sun is high in the sky; half a year later, in winter, it is low.

We measure time by Earth's movements relative to the Sun. One complete spin of Earth on its axis is a day. Earth spins 365¼ times before it returns to the same point in its orbit around the Sun; this is the year. The odd quarter day builds up so every fourth year is a "leap" year to keep the calendar in step with the seasons.

13

Land, sea, and air

The continents are enormous rafts of rock adrift on a molten sea, pushed over the face of the Earth by the slow-moving currents of global convection.

To a visitor from another star system, Earth would stand out among the planets orbiting our Sun because from space it looks as if it is almost entirely covered in water. But the oceans are not just low-lying areas that happen to be water-filled. Even if there were no water, there would still be a clear difference between the appearance and activity of the upland areas and the ocean troughs.

At the centers of the ocean areas are vast volcanic ridges, where a type of rock called basalt constantly wells up, creating huge "plates" of material. Driven by the convection currents of material rising from Earth's molten interior, the rock slowly moves outward from these mid-ocean ridges, causing sea-floor spreading and pushing the plates along. The plates move over the underlying material at a rate of an inch or two a year.

Earth's continents are giant rafts up to 100 miles (160 km) thick sitting on top of the plates. Some 200 million years ago, there was just one landmass, but it broke up to form today's continents. Where plates have collided, vast mountain ranges such as the Himalayas and Alps have arisen; where they slip past each other, a series of earthquakes occurs; where one plate disappears beneath another, deep ocean trenches form and volcanic activity takes place.

Gases belched out during volcanic eruptions over millions of years gave Earth its initial atmosphere, which was changed by geological action and by plants converting carbon dioxide to oxygen. The make-up of air has settled at 77 percent nitrogen, 21 percent oxygen, 1 percent water vapor, and smaller amounts of argon, carbon dioxide, neon, helium, and sulfur. Earth's gravitational field is strong enough to stop the atmosphere from leaking away into space.

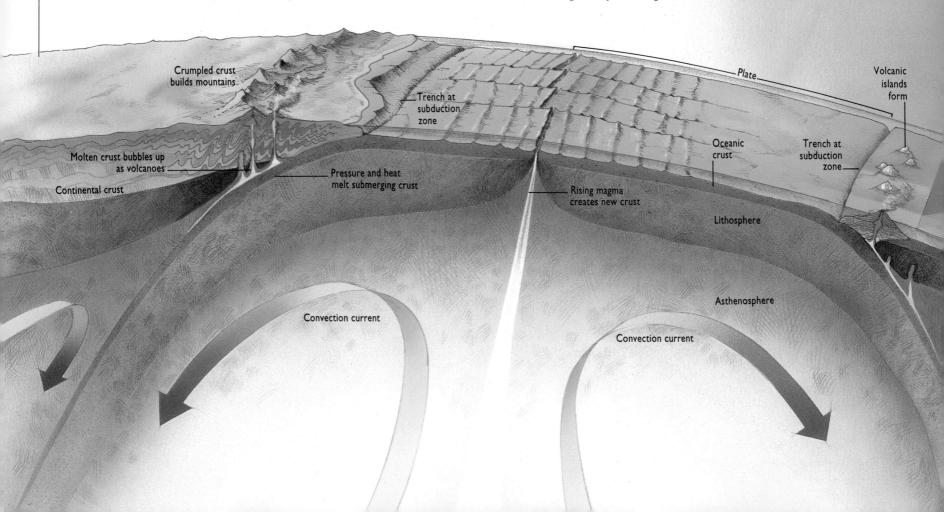

Crumpled crust builds mountains

Trench at subduction zone

Plate

Volcanic islands form

Molten crust bubbles up as volcanoes

Continental crust

Pressure and heat melt submerging crust

Oceanic crust

Trench at subduction zone

Rising magma creates new crust

Lithosphere

Asthenosphere

Convection current

Convection current

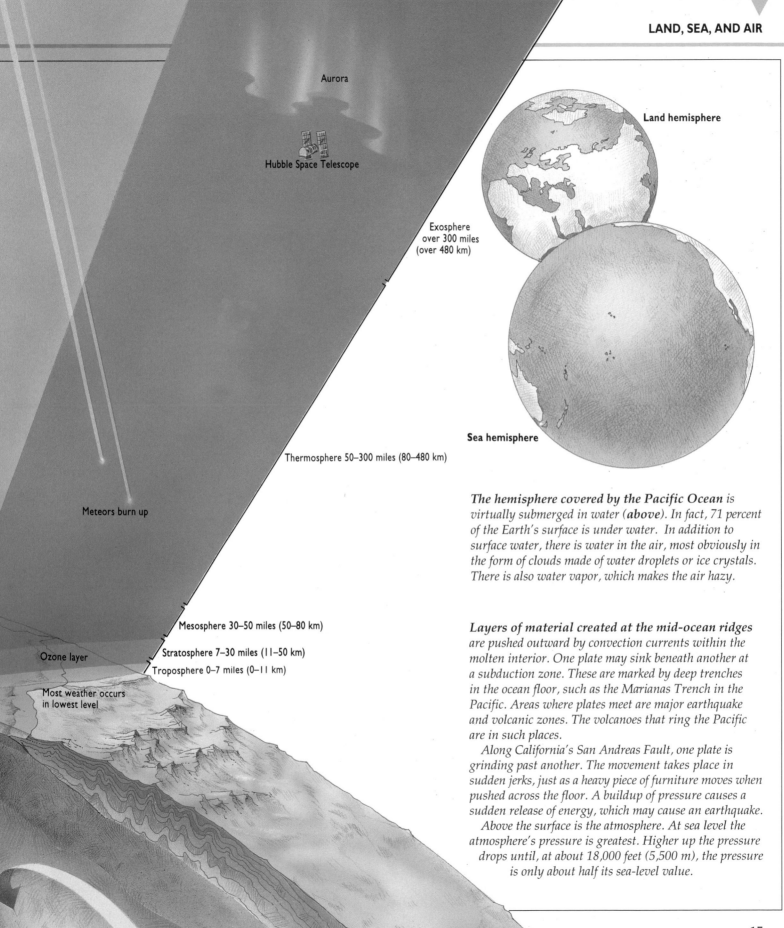

Aurora

Hubble Space Telescope

Land hemisphere

Exosphere
over 300 miles
(over 480 km)

Sea hemisphere

Thermosphere 50–300 miles (80–480 km)

Meteors burn up

Mesosphere 30–50 miles (50–80 km)

Stratosphere 7–30 miles (11–50 km)

Ozone layer

Troposphere 0–7 miles (0–11 km)

Most weather occurs
in lowest level

*The hemisphere covered by the Pacific Ocean is
virtually submerged in water (above). In fact, 71 percent
of the Earth's surface is under water. In addition to
surface water, there is water in the air, most obviously in
the form of clouds made of water droplets or ice crystals.
There is also water vapor, which makes the air hazy.*

*Layers of material created at the mid-ocean ridges
are pushed outward by convection currents within the
molten interior. One plate may sink beneath another at
a subduction zone. These are marked by deep trenches
in the ocean floor, such as the Marianas Trench in the
Pacific. Areas where plates meet are major earthquake
and volcanic zones. The volcanoes that ring the Pacific
are in such places.*

*Along California's San Andreas Fault, one plate is
grinding past another. The movement takes place in
sudden jerks, just as a heavy piece of furniture moves when
pushed across the floor. A buildup of pressure causes a
sudden release of energy, which may cause an earthquake.*

*Above the surface is the atmosphere. At sea level the
atmosphere's pressure is greatest. Higher up the pressure
drops until, at about 18,000 feet (5,500 m), the pressure
is only about half its sea-level value.*

Origins of life

The rich diversity of life on Earth became possible when chance threw together a combination of chemicals in a structure that could copy itself.

Our world seethes with life. Living things are found almost everywhere on our planet's surface, from the lush forests of the tropics to the extraordinary "dry valleys" of Antarctica, where hardly any moisture has fallen in thousands of years. The most sterile operating room contains microscopic bacteria, and ordinary houses shelter millions of living things, from tiny fungi to mites and insects. But how did all this life first appear?

For many people, the existence of life on Earth can be explained only by a special act of creation. But evidence from fossils suggests that life has existed on Earth for billions of years and has gradually changed and developed as time has gone by. When the Earth was young – over four billion years ago – it was too hot to support life, and its surface was bombarded by meteors hurtling in from space. Gradually Earth cooled and developed a skin of water. It is in this global ocean that life probably began.

The ancient ocean was a huge vat of simple chemical compounds. The atoms that made up these compounds constantly changed partners. Sometimes atoms joined to form bigger, more complex compounds. But sooner or later, these broke up, and vanished without trace.

Life became possible only when an extraordinary chance threw together a compound that could make copies of itself. Like a chain of magnets, it attracted its own chemical components, lining them up to form a new chain. Then the new chain broke free and, in turn, made copies of itself.

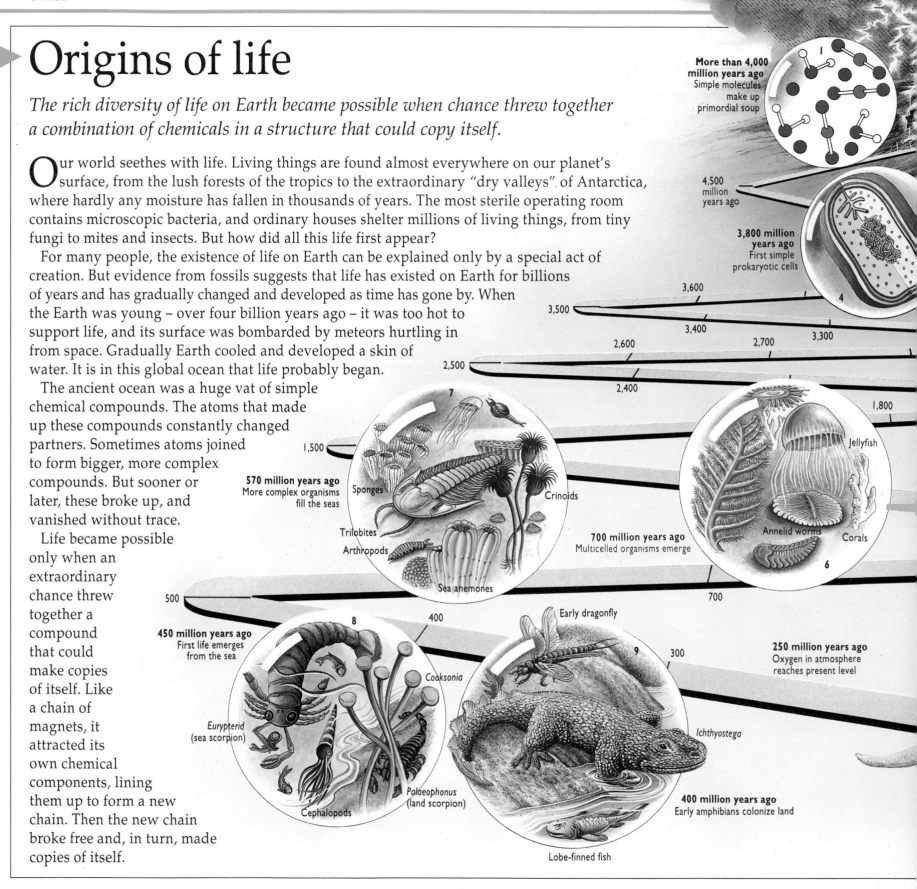

More than 4,000 million years ago
Simple molecules make up primordial soup

4,500 million years ago

3,800 million years ago
First simple prokaryotic cells

3,600

3,500

3,400

3,300

2,700

2,600

2,500

2,400

1,800

1,500

570 million years ago
More complex organisms fill the seas

Sponges

Crinoids

Jellyfish

Trilobites

Arthropods

Annelid worms

Corals

Sea anemones

700 million years ago
Multicelled organisms emerge

700

500

400

Early dragonfly

450 million years ago
First life emerges from the sea

300

Cooksonia

250 million years ago
Oxygen in atmosphere reaches present level

Eurypterid (sea scorpion)

Ichthyostega

Palaeophonus (land scorpion)

Cephalopods

400 million years ago
Early amphibians colonize land

Lobe-finned fish

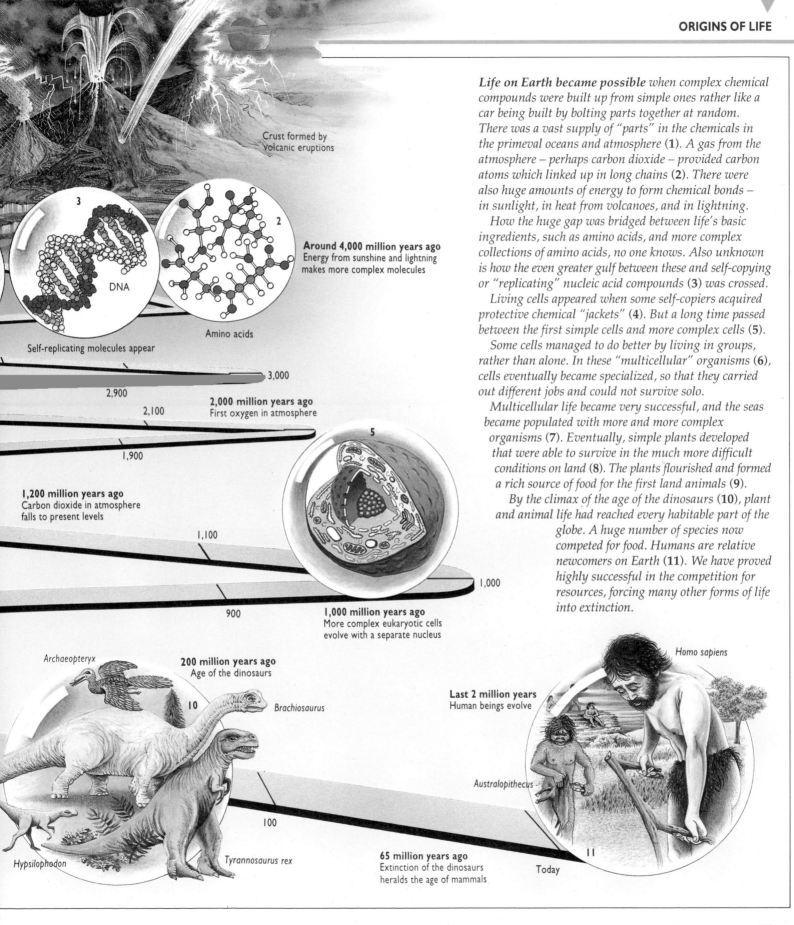

Crust formed by volcanic eruptions

Around 4,000 million years ago
Energy from sunshine and lightning makes more complex molecules

DNA

Amino acids

Self-replicating molecules appear

3,000

2,900

2,100

1,900

2,000 million years ago
First oxygen in atmosphere

1,200 million years ago
Carbon dioxide in atmosphere falls to present levels

1,100

1,000

900

1,000 million years ago
More complex eukaryotic cells evolve with a separate nucleus

Archaeopteryx

200 million years ago
Age of the dinosaurs

10

Brachiosaurus

Hypsilophodon

Tyrannosaurus rex

100

65 million years ago
Extinction of the dinosaurs heralds the age of mammals

Today

Homo sapiens

Last 2 million years
Human beings evolve

Australopithecus

11

Life on Earth became possible when complex chemical compounds were built up from simple ones rather like a car being built by bolting parts together at random. There was a vast supply of "parts" in the chemicals in the primeval oceans and atmosphere (**1**). A gas from the atmosphere – perhaps carbon dioxide – provided carbon atoms which linked up in long chains (**2**). There were also huge amounts of energy to form chemical bonds – in sunlight, in heat from volcanoes, and in lightning.

How the huge gap was bridged between life's basic ingredients, such as amino acids, and more complex collections of amino acids, no one knows. Also unknown is how the even greater gulf between these and self-copying or "replicating" nucleic acid compounds (**3**) was crossed.

Living cells appeared when some self-copiers acquired protective chemical "jackets" (**4**). But a long time passed between the first simple cells and more complex cells (**5**).

Some cells managed to do better by living in groups, rather than alone. In these "multicellular" organisms (**6**), cells eventually became specialized, so that they carried out different jobs and could not survive solo.

Multicellular life became very successful, and the seas became populated with more and more complex organisms (**7**). Eventually, simple plants developed that were able to survive in the much more difficult conditions on land (**8**). The plants flourished and formed a rich source of food for the first land animals (**9**).

By the climax of the age of the dinosaurs (**10**), plant and animal life had reached every habitable part of the globe. A huge number of species now competed for food. Humans are relative newcomers on Earth (**11**). We have proved highly successful in the competition for resources, forcing many other forms of life into extinction.

The struggle for survival

Nothing in nature is constant. Over millions of years, the processes of evolution have shaped all living things, including people.

Imagine that we could somehow travel backward in time, just 50 or 100 years into the past. The human world, with its strange fashions and clumsy technology, would seem very different to that of today. The natural world, on the other hand, would be reassuringly familiar. But if we could go back much farther than this, not one century but a million, we would find a world totally unlike our own. Not only would humans and most other mammals be missing, but the landscape would be covered by strange plants and dominated by awe-inspiring reptiles – the dinosaurs.

True time travel is still beyond our grasp. However, by studying fossils – preserved replicas in stone of long-dead plants and animals – a picture of the past can be built up. This fossil record shows that instead of staying the same, living things have gradually changed, or "evolved," over time.

In his book *On The Origin of Species*, published in 1859, evolution pioneer Charles Darwin tried to explain how this process of change occurs. He knew that living things vary from each other and that characteristics are passed on from parent to offspring. Living things can reproduce very quickly until resources, such as food, become limited. When this happens, all life forms must struggle to survive,

When a lion attacks a herd of zebras (below), it is playing an unconscious part in the never-ending weeding-out process which guarantees that all living things continue to evolve. Any young zebra with a weakness such as poor eyesight is more likely to be caught by a lion. By killing the zebra, the lion prevents it from passing the same flaw down to the next generation.

Although the zebra's stripes seem to make a lone animal an easy target for the lion, they appear to have evolved with a dual purpose. The distinctive markings stimulate zebras to graze together in herds. In these groups, the confusing mass of stripes stops predators from identifying individual animals.

a competition both between species and within them. The prize for those that adapt best to the challenge is successful breeding.

Ways of gaining advantage are as varied as nature itself, but include living with other species, finding new sources of food, and fooling predators with camouflage.

This ever-continuing process of weeding out the weakest is known as natural selection. Though refined since Darwin's day, this theory remains the cornerstone of evolution.

Giraffes evolved long necks (left) because those with slightly longer necks were able to reach more food and so raise more young. As this process repeated itself, average neck length got longer. The trend stopped when the disadvantages – such as pumping blood through a longer neck – began to outweigh the advantages.

Despite their variety of colors, these brittle stars (below) are all members of the same species. This kind of variety, called polymorphism, happens when natural selection favors no single characteristic, such as color, above any other. If brittle stars of one color ever began breeding more successfully, they would soon dominate.

A place to live

*Earth is the only place where we know for sure that life exists.
So what are the factors that make our planet so special?*

The Earth abounds with living things, but its nearest neighbor, the Moon, is barren and lifeless. Earth and Moon are about the same distance from the Sun, and they both receive energy in the form of sunlight, so why is one fertile and the other biologically dead? One of the reasons Earth is able to support life has to do with its size. The Earth is larger and also about 80 times heavier than the Moon, and so it exerts a much stronger gravitational pull than its junior partner. This pull holds the atmosphere in place, and the average temperature at the Earth's surface of 59°F (15°C) means that water at the surface is in liquid form. Water and water vapor and other atmospheric gases play a crucial role in creating the right conditions for living things to flourish.

The atmosphere acts both to screen and insulate the Earth. The energy the Sun emits is in the form of a broad band of radiation. The Earth's atmosphere is almost transparent to visible light from the Sun, but it blocks much of the Sun's ultraviolet light and all its X-rays, which can break up the substances that form living tissue. By acting as a screen, the atmosphere shields life on Earth from this kind of damage.

By acting as an insulator, the atmosphere moderates temperature changes. On the Moon, where there is no atmosphere, the surface temperature soars to 215°F (102°C) after the Sun rises and then plunges to –251°F (–157°C) after it sets. On Earth, temperatures also rise and fall, but by much smaller amounts. This is because the atmosphere both screens incoming radiation and retains some of the heat radiated back by the ground. The Earth's surface temperature is also steadied by the oceans, which are slow to heat up and slow to cool down.

This combination of abundant water and a protective atmosphere is found on no other planet orbiting our Sun. But the fact that Earth is unique in the solar system does not mean it is unique in our galaxy. Many astronomers now believe that planetary systems are not as rare as they were once thought to be. If just a few of the planets orbiting distant stars are similar to Earth, the number that could have conditions suitable for life may run into many millions. Whether life has actually arisen on any of them, and whether it has attained an intelligent form, are questions that have yet to be answered.

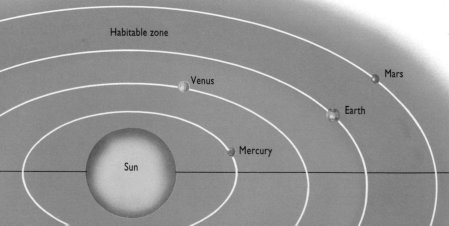

Habitable zone

Venus

Mars

Earth

Mercury

Sun

EARTH CALLING

In an attempt to contact any intelligent life in the Galaxy, a message was beamed from the radio-telescope at Arecibo, Puerto Rico, in 1972. It told of the existence of life on Earth in a simple, easy-to-decipher binary code. It first set out the numbers from 1 to 10, then depicted the key chemicals that make up living things. An image of a human and of the telescope completed the message. So far, there has been no reply.

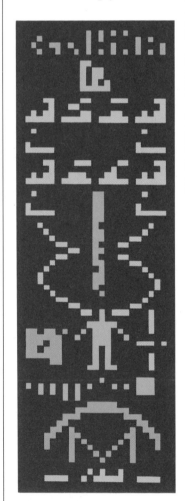

There is around the Sun a zone in which its energy, if received by a suitable planet, is sufficient to keep the temperature at the right levels for water to be liquid at the surface of the planet (**left**, not to scale).

Earth orbits the Sun comfortably within this zone. Over most of its surface, water is found in liquid form and conditions are suitable for life (**above**). On Venus and Mars, surface conditions do not allow

water to be liquid. On Venus it is too hot because it is too close to the Sun and because there is a runaway greenhouse effect. On Mars it is too cold both because the planet is further from the Sun and because Mars is not massive enough to have the gravity to keep its insulating atmosphere. Temperatures rise above the freezing point of water during the Martian summer, but no signs of life have been detected.

Our local star

The Sun is center of the solar system and our main source of light and heat – but it is just an ordinary star.

The Sun is like millions of other stars in the universe. Since we are so close to it – a mere 93 million miles (149,600,000 km) – we can study it closely. It may look solid, but like all stars, it is just a ball of gas consisting mostly of hydrogen, but with a fair proportion of helium. And even though it is 109 times as wide as the Earth, it is small by star standards, with a diameter of 865,300 miles (1,392,500 km).

The Sun's core is incredibly hot – around 27 million °F (15 million °C) – and temperatures like these strip all the atoms of hydrogen to the bare nucleus. So the center of the Sun is a dense, jostling sea of minute nuclear particles called protons. Protons usually repel one another, like identical magnetic poles, but at such extremely high temperatures and pressures they may occasionally be forced together, a process called nuclear fusion. Such fusions are rare events, but there are so many protons in the Sun's core that there they happen all the time.

When two protons are forced together, they start a chain of events which eventually changes them into helium and releases a phenomenal amount of energy. This energy is mainly short-wavelength gamma radiation and X-rays, which are much more energetic than radiation at the wavelengths of visible light. Nuclei of gas atoms at the Sun's core absorb and then re-emit much of this radiation as it moves outward.

It takes tens of thousands of years for energy generated in the core to reach the Sun's surface. Some 375,000 miles (600,000 km) from the Sun's center, the gas is much cooler. Normal hydrogen and helium atoms can form, and energy then rises, mainly as light and heat. The Sun's surface, or photosphere – which is a white disk in the sky – shines at a modest 10,500°F (5,800°C).

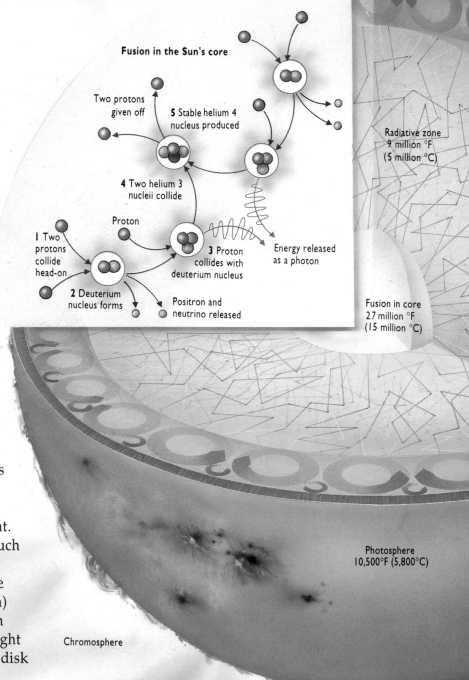

Fusion in the Sun's core

Two protons given off

5 Stable helium 4 nucleus produced

4 Two helium 3 nucleii collide

Proton

1 Two protons collide head-on

3 Proton collides with deuterium nucleus

Energy released as a photon

2 Deuterium nucleus forms

Positron and neutrino released

Radiative zone 9 million °F (5 million °C)

Fusion in core 27 million °F (15 million °C)

Photosphere 10,500°F (5,800°C)

Chromosphere

At the Sun's heart *is a dense core of hydrogen and helium nuclei. When two hydrogen nuclei fuse, they make heavy hydrogen or deuterium. With the addition of a proton, helium 3 is formed, and a massive amount of energy released. When two helium 3 nuclei collide, a* stable nucleus of helium 4 is made. The energy released gradually radiates up through the "radiation zone." As this energy nears the surface, it moves up through the "convection zone" as masses of hot gas. These rise to the surface, cool, and sink, mottling the photosphere with cells of bright rising gas ringed by darker sinking gas. In some places, strong magnetic fields cut down the light output. This creates cooler, darker areas called sunspots.

The photosphere is usually so bright that the chromosphere, the pinkish layer of gas beyond, can only be seen during a solar eclipse. In places, "prominences" of hot hydrogen gush many thousands of miles high. Beyond the chromosphere is a ring or corona of gases at temperatures of millions of degrees – but with little real heat, since the gas is so rarefied.

Prominences

Convection zone

Sunspots

Solar flare

The Sun is called a yellow star *because its peak output is yellow light, and most people think of it as yellow. In fact, the Sun is visually pure white. It looks yellow or red at sunset because the atmosphere absorbs bluer light in the spectrum when the Sun is low in the sky.*

The giver of life

Virtually everything that lives on Earth depends on energy from our local star, the Sun. Solar power also drives the weather systems that bring clouds, wind, and rain.

Earth's powerhouse is the Sun. If it shut down, our planet would become a dark, dead, frozen lump of rock. When the Sun shines, we feel its power as warmth on our skin, but the Sun's effects are more profound – it is the power behind life itself. The Sun gives energy to a chemical process inside plants – photosynthesis – which converts the Sun's raw energy into glucose, a source of energy that fuels the food chain, and on which virtually all life on Earth depends. We use the remains of plants that converted sunshine millions of years ago as fossil fuels – coal and oil. In fact, all our energy, except for nuclear, geothermal, and tidal power, comes from the Sun.

The Sun pumps out energy at an enormous rate. Even as far away from the Sun as the Earth's orbit – an average of 93 million miles (150 million km) – 1.2 kilowatts per square yard (1.4 kilowatts/m^2) of energy arrive in space. Earth's atmosphere absorbs and reflects some of this energy so at the surface, with the Sun directly overhead, each square yard receives nearly a kilowatt – the same as the light from ten 100-watt bulbs.

The equator heats up more than the poles because the Earth is round, and at higher latitudes the Sun's energy is spread over a larger area than at the equator. Weather is a system that redistributes heat away from the hotter regions. The Sun drives the wind and evaporates water, which later falls as rain or snow in higher latitudes.

All Earth's weather *systems are part of a giant heat-exchange machine driven by the Sun. This machine moves excess heat from the equator and spreads it more evenly about the globe. Sunshine at the equator heats the surface more than at other latitudes, where the Sun's energy is spread over a larger area.*

At the equator the air above the surface gets hot, rises, and flows to the tropics, leaving a permanent area of low pressure. Where this heated air sinks, there is a high-pressure area. Winds are caused by air flowing back across the surface from high to low-pressure regions. This Sun-driven pattern of air-circulation cells is repeated at different latitudes.

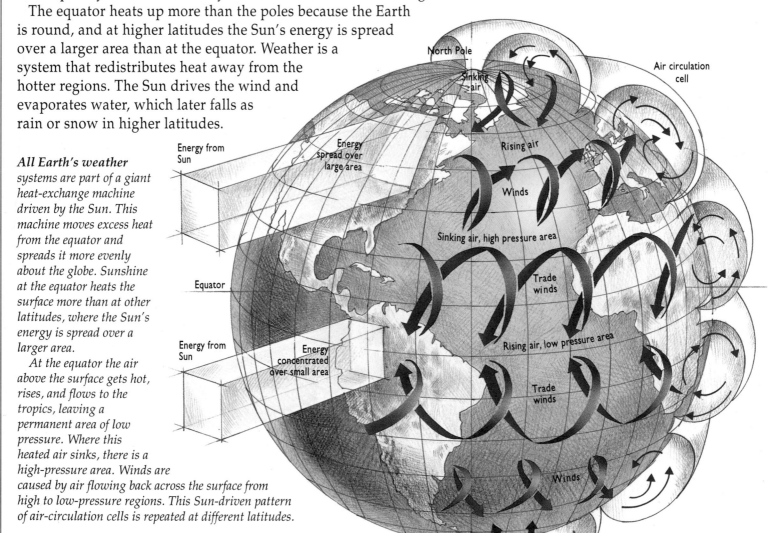

North Pole

Sinking air

Air circulation cell

Energy from Sun

Energy spread over large area

Rising air

Winds

Sinking air, high pressure area

Equator

Trade winds

Energy from Sun

Energy concentrated over small area

Rising air, low pressure area

Trade winds

Winds

Bubbling cloud *shows the Sun's power at work (**left**). As the Sun beats down on water, evaporation takes place. The warm, moist air rises and meets cooler air, and the vapor condenses to form tiny water droplets suspended in the air as cloud. Swirling air currents within clouds make millions of these droplets collide and join together into raindrops. When they are heavy enough, the raindrops fall from the cloud to replenish our vital fresh water supply.*

The Sun's power *could potentially fulfill all our energy needs – the problem is collecting and storing it. This solar furnace (**right**) in Odeillo, France, is used to research ways of doing just that. Flat mirrors track the Sun across the sky, reflecting it onto a huge, curved mirror, which then focuses the energy onto part of the tower in front. This super-concentrated sunshine heats water to make steam which drives an electric generator.*

The shimmering, ethereal displays *that illuminate the polar skies are the aurorae, or the northern or southern lights. Charged particles from the Sun stream toward the Earth at immense speeds. The* *particles are channeled to the poles by Earth's magnetic field and hit the upper atmosphere. Here they excite air molecules until they glow, making arcs and streams of glowing yellow, green, and red.*

Earth's rocky neighbors

The internal structures of the four inner planets have a lot in common. But on the surface it is an entirely different matter.

	I Mercury
Diameter miles (km)	3,031 (4,878)
Distance from Sun million miles (million km)	36 (58)
Number of satellites	0
Length of day (Earth days taken to rotate once)	58.6
Year (Earth days to orbit Sun)	88

The four planets closest to the Sun are similar enough to be known as the terrestrial (Earth-like) planets because, like Earth, they are made of rock. But Mercury, Venus, Earth, and Mars are all unique. The size and distance of each planet from the Sun have shaped the surface characteristics of each one.

The smaller the planet, the harder it has been for it to keep its atmosphere in the glare of the Sun. So tiny Mercury, closest to the Sun, never had a chance of keeping its atmosphere. Venus was large enough to keep its atmosphere, but too close to the Sun and too hot for water to remain liquid. Mars, small and the farthest from the Sun of the terrestrial planets, is cold and desolate with a thin atmosphere. Only Earth is the right size and distance from the Sun to retain enough atmosphere to create the correct temperature for liquid water and life.

Crust — Core — Mantle

Mercury has a silicate crust and, like Earth, a dense interior with a nickel-iron core. Unlike Earth, however, there is little evidence of volcanoes, probably because Mercury cooled down quickly after it formed. But the planet is not totally inactive, and there are hot regions of volcanic activity below the surface, known as hot poles. Mercury has a fairly elliptical orbit, and when it is closest to the Sun, the surface temperature can reach 873°F (467°C). This intense heat and the planet's weak gravitational field mean that any atmosphere it had boiled away long ago. Because of this, the planet is crater-pitted, rather like our moon, because it has no atmosphere to burn up lumps of debris that collide with it and no erosion to wear down the craters formed after impacts. There are traces of hydrogen and helium, but too little to affect the surface conditions in any way.

Clouds of sulfuric acid — Core — Mantle — Crust

Venus is almost exactly the same size and mass as Earth, with a similar interior, including a nickel-iron core. But unlike Earth, Venus is unbearably hot and covered with thick, choking gases. Huge volcanoes, which may still be active, gave it a dense atmosphere dominated by carbon dioxide – which now makes up 96 percent of the atmosphere – with some nitrogen and traces of other gases. Its orbit takes it closer to the Sun, and it receives twice as much solar radiation as Earth. The deep carbon dioxide atmosphere causes a powerful greenhouse effect which traps this extra heat, creating Venus's high surface temperature. Earth's average surface temperature is 59°F (15°C) while that of Venus is 855°F (457°C) – above the melting point of lead. To complete this inhospitable picture, Venus has clouds made of poisonous sulfuric acid, a substance which also falls as a deadly rain.

2 Venus	3 Earth	4 Mars
7,520 (12,102)	7,926 (12,756)	4,218 (6,786)
67 (108)	93 (150)	142 (228)
0	1	2
243	1	1.03
225	365	687

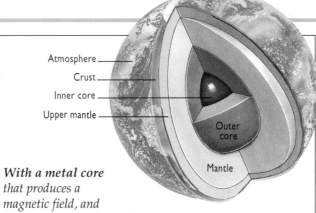

Atmosphere

Crust

Inner core

Upper mantle

Outer core

Mantle

With a metal core

that produces a
magnetic field, and
volcanic activity to provide an
atmosphere, Earth is a typical inner planet. The surface
temperature is maintained because the atmosphere acts
like a blanket. It insulates the Earth, keeping the surface
at a temperature where water is a liquid. This has been
crucial to the way our world has developed. Eons ago,
green plants evolved in water and took carbon dioxide
from the air, replacing it with oxygen. Our air today is
mainly nitrogen and oxygen with some water vapor.
There are also trace gases, including carbon dioxide.

Earth's core is nickel-iron with a metallic center and
a liquid outer core. Above the core is the mantle, which
is solid and rocky, although the upper mantle, or
asthenosphere, is more like thick heavy molasses with
a thin and brittle outer crust.

Mars, the red planet, is

actually colored red by
rust – iron oxide
and other iron
impurities that
cover the whole
surface. Despite
the rust on its
surface, Mars has
only a small iron
core and is less
dense than the other
inner worlds. Volcanoes
on its surface gave Mars its
thin atmosphere of carbon dioxide,
which also makes up its white ice caps. In fact, the
Martian atmosphere is composed mainly of carbon
dioxide (95 percent) although there are small amounts
of nitrogen, oxygen, and water vapor.

Mars now has a much thinner atmosphere than either
Venus or Earth. But the presence of apparently dried-up
water channels on its surface suggests that its atmosphere
was once thick enough for rain to fall. With an air
temperature well below freezing, and high winds that
whip the sand up into frequent dust storms, Mars is a
hostile place, and there are no signs that life has evolved.

Crust

Core

Mantle

Great balls of gas

Huge gas storms and majestic rings typify the power and the beauty of the huge outer planets.

	5 Jupiter	6 Saturn
Diameter miles (km)	88,849 (142,984)	74,900 (120,536)
Distance from Sun million miles (million km)	483 (778)	887 (1,427)
Number of satellites	16	18
Length of day (Hours to rotate once)	9.8	10.2
Length of year (Earth years to orbit Sun)	11.9	29.5

People may someday set foot on each of the inner planets – but they will never be able to land on the four giant ringed outer planets. Jupiter, Saturn, Uranus, and Neptune all have solid cores, but are far enough from the Sun to have kept their thick, dense, and dangerous layers of gas. Of the five outer planets Pluto, farthest from the Sun, is the odd one out. It is a small ball of ice, more like an escaped asteroid than a planet.

Rocky core

Metallic hydrogen

Great Red Spot

Hydrogen gas

Circulation cells of liquid hydrogen

If Jupiter, the largest planet of the solar system, were just a little more massive, it would have become a small star. Although Jupiter's core is as hot as the surface of the Sun, it never reached the temperatures needed to start the nuclear reactions that make stars shine. Above the planet's hot, rocky heart is a layer of hydrogen subjected to such huge pressure that it behaves like a metal. Much of the planet is, in fact, made of hydrogen in various states. Jupiter's vivid surface colors are produced by traces of methane, phosphorus, and ammonia. The pattern of bands – caused by the different layers in the planet – is always changing. But one permanent feature is the Great Red Spot – a huge storm colored by phosphorus which reddens in the Sun's ultraviolet light.

Rocky core

Metallic hydrogen and helium droplets

Circulation cells of hydrogen and helium gas

Saturn, the sixth planet of the solar system, would be merely a pale version of Jupiter were it not for its glorious ring system. Though all the gas giants have rings, only those of Saturn can be seen directly from Earth through a telescope. There are more than 10,000 individual rings, each composed of countless tiny ice particles, all of which orbit the planet in bands no more than ⅔ mile (1 km) thick. Apart from its rings, Saturn is very similar to Jupiter, with surface streams of fast-moving cloud giving an ever-changing pattern of spots and eddies. The interior is also like Jupiter, though much less dense – in fact, Saturn is lighter even than water and would float if given the chance. This lightness, together with the planet's fast speed of rotation, means that Saturn, more than any other planet, bulges at its equator and is rather flattened at its poles – a shape known as oblate.

5

7 Uranus	8 Neptune	9 Pluto
31,764 (51,118)	30,776 (49,528)	1,429 (2,300)
1,784 (2,871)	2,794 (4,497)	3,675 (5,914)
15	8	1
17.9	19.2	6.4
84	165	248.5

6
7
8
9

Atmosphere of hydrogen, helium, and methane

Mantle of frozen water, methane, and ammonia

At the heart of Uranus there probably lies a rocky core, but most of the planet is made up of ammonia, water, and a thick greenish atmosphere of hydrogen, helium, and methane. What distinguishes Uranus is the tilt of its axis. The angle is so great that the planet rotates the opposite way to all other planets. This means that its poles, usually the coldest parts of a planet, face the Sun and can actually be warmer than its equator.

Mantle of frozen water and ammonia

Rocky core

Atmosphere of hydrogen, helium, and methane

Neptune is the most distant of the gas giants. Its core is probably not solid; it is more likely to be a mixture of ice and rock, which gives way to ice and gas. Methane gas in large amounts gives the planet its blue color. Methane also adds wispy white clouds to the planet's atmosphere. Parts of Neptune's three rings appear brighter than others due to the uneven spread of particles.

Pluto is not a gas giant but a tiny icy body, smaller than our moon. For most of its journey around the Sun, Pluto is the most remote of the solar system's planets. However, its elliptical orbit occasionally brings Pluto closer to the Sun than Neptune. Pluto's strange orbit and tiny size mean that it is probably not a real planet, but a large asteroid which may have escaped from another orbit. Pluto has its own satellite, close-orbiting Charon, which is just about half the diameter of Pluto itself.

Methane frost

Rocky core

Mantle of ices

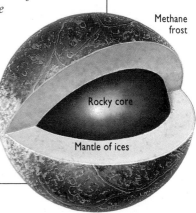

Planetary attendants

*Some of the most fascinating and varied worlds in the solar system
are the satellites in orbit around the planets.*

The satellites, or moons, of the planets come in a huge variety. Jupiter's satellite Callisto looks like an icy pond that has refrozen after being shattered, while Uranus's Miranda boasts ice cliffs which could be a tourist venue of the future. Like the planets around which they orbit, satellites appear to form two distinct groups – the rocky, higher-density ones from Mercury to Mars, then ice rock mixtures from Jupiter to the outer solar system. For example, the density of our moon is 3.34 times that of water, Mars's satellite Phobos is 2.2, similar to volcanic rock, while Saturn's Mimas is 1.17, slightly denser than water ice (0.9).

One of the biggest satellites in the solar system is Saturn's Titan. It is the only satellite with an atmosphere and it seems to be partly covered by an ocean, probably of liquid ethane, though this cannot be seen through the clouds. An organic haze both blocks the view of the surface and stops sunlight from reaching Titan's surface.

Thebe

Adrastea

Metis

Amalthea

Callisto

Ganymede

Europa

Io

Jupiter's path
around the Sun

Elara

Himalia

Lysithea

Leda

Ananke

Carme

Sinope

Pasiphaë

Most of the lunar features visible with the naked eye
are the result of impacts with space debris. Current
thinking is that the Moon itself was born from the Earth
at an early stage, torn out of it in a huge splash of molten
rock following a collision with a chunk of protoplanet.

Giant planet Jupiter's gravity holds 16 known
satellites in orbit (**left**). The inner ones, including the
four visible through binoculars (Io, Europa, Ganymede,
and Callisto), orbit directly above Jupiter's equator. They
probably formed at the same time as Jupiter itself, much
as the planets formed around the Sun. The outer ones are
in more erratic orbits and are probably captured debris.
Some of the outer ones are no more than 12 miles (20 km)
across. Io (**right**), bigger than our own Moon, is so close
to Jupiter that the planet's gravitational field churns up
Io's interior. As a result, it has volcanoes, one of which is
seen here as a plume.

Space debris

There is more to the solar system than Sun, planets, and satellites. Leftovers from planet formation are some of the most spectacular, and dangerous, bodies in space.

Beyond the inner planets is a ring of rocky debris orbiting the Sun between Mars and Jupiter. Some of this debris is big – the largest piece, called Ceres, is about 620 miles (1,000 km) wide; but most of the 5,000 or so known pieces are only a few miles across. These are the asteroids, or minor planets.

At the edge of the known solar system, beyond the gas planets, there is thought to lie yet more debris, this time mostly ice rather than rock. From time to time, the inner solar system receives visitors from these outer reaches in the form of comets. Comets are like large, dirty snowballs. When their orbit brings them close to the Sun, the surface is vaporized, and dust mixed in with the ice is blown off the surface to trail behind, shining with reflected sunlight.

When Earth hits a comet trail, particles of the fragile, crumbly dust burn up in our atmosphere because of friction energy generated as they hit air molecules. The brief streak of light emitted as they burn is called a shooting star or meteor.

About 100,000 years ago, a bit of rocky debris over 100 feet (30 m) across hit the Earth at high speed, creating Meteor Crater in Arizona. If a similar object were on course for Earth now, there would be little warning; even if there were, there is not much we could do – the impact could be as powerful as a nuclear explosion. Many scientists believe a major impact 65 million years ago threw so much dust into the atmosphere that the climate altered, causing the death of the dinosaurs and many other species.

Most asteroids orbit the Sun in a belt between Mars and Jupiter. They do not occupy certain orbits where Jupiter's gravity disturbs them, leaving what are called Kirkwood Gaps. Jupiter's gravity has also corraled two groups of asteroids, called the Trojans, so that they share the planet's orbit, but are 60° in front of and behind Jupiter.

Some asteroids, such as Icarus, travel closer to the Sun than even Mercury, while others, like Hidalgo, orbit beyond Jupiter. Most of the asteroids orbit in the same plane as the planets, but comets can be on much more inclined orbits. New comets may have open-ended orbits and enter the inner solar system only once, while others become trapped by gravity within the solar system and have orbits that regularly come close to the Sun and then swing out again.

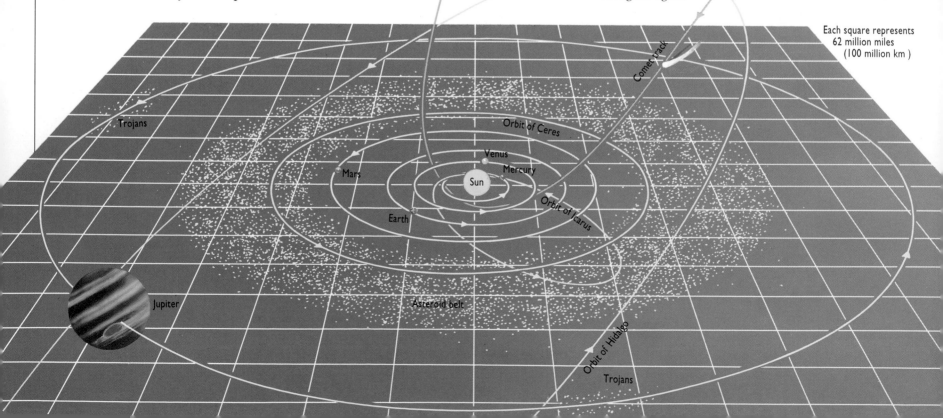

Each square represents 62 million miles (100 million km)

Trojans

Comet track

Orbit of Ceres

Venus

Mercury

Mars

Sun

Orbit of Icarus

Earth

Jupiter

Asteroid belt

Orbit of Hidalgo

Trojans

***Comet West* (left)** put on this dramatic display in March, 1976. Big comets such as West are rare; most look like dim, fuzzy patches of light. But a big comet can have a brilliant head visible in daylight and a tail that stretches as far as the distance between Earth and Sun.

The head and tail of a big comet are insubstantial, yet spectacular, displays of glowing, scattered particles. The solid body of a comet is just an icy lump only a few miles across. The display happens because, as the comet nears the Sun, its ice vaporizes and grains of rock held in the ice are released and reflect the Sun's light.

The tail of a comet is made of gas and dust. The gas tail of Comet West shows clearly, fluorescing blue in light from the Sun. The gas tail is blown out straight by the solar wind – a stream of particles from the Sun. The white dust tail is curved because it is left behind by the comet as it orbits. Both tails always point away from the Sun, despite the comet's orbital motion.

Gaspra is a typical asteroid. Its small size – 12 miles (19 km) long – means that its internal gravity is too small to pull it into a sphere.

Although tiny, Gaspra compares in size with Manhattan Island, New York (**below**). Craters on Gaspra show that it has suffered many collisions.

Manhattan

Gaspra

33

Central bulge of old stars

Systems of stars

Orbiting the center of the galaxy, our Sun is in the thick of a busy part of the stellar neighborhood.

The Sun may be an average star, but at least it is in a pretty exciting part of town – or, rather, part of the galaxy. The galaxy is our own star-city, made up of about 100 billion other stars in a spiral structure. We can look out of our galaxy to see millions of other galaxies throughout the universe – possibly 100 billion galaxies, maybe more.

What we call the Milky Way is the galaxy seen from inside. Except in remote places, most people do not get to see it at all these days because of light pollution from streetlights, but on a dark night it can look stunning. Our location on Earth in one of the spiral arms gives a grandstand view. There are glowing gas clouds where stars are born, as well as clusters of new stars and dark dust lanes. All these are signs that the galaxy is youthful and thriving, with stars being created all the time.

The Orion region of our galaxy, our nearest major star-birth area, contains many clouds of gas or dust, called nebulae. In photographs, parts of this region show spectacular colors. For instance, there is pink hydrogen gas fluorescing in the light from a bright star nearby. In addition to gas, there is dust, some of which appears blue in the Orion nebula. The dust does not emit light, but reflects the light of stars embedded within it. The dust scatters the light, just as the molecules of our atmosphere scatter sunlight and make our sky blue. Dusty regions such as these can fragment into smaller patches which begin to shrink and form stars.

To see other galaxies, we must look sideways, at right angles from the Milky Way, where there are fewer stars and dust clouds to hinder our view. There we see remote patches of light which may turn out to be spiral galaxies like our own, elliptical galaxies which lack spiral arms, and young stars, or a variety of irregular galaxies in all shapes and sizes. Galaxies like company: there are clusters of maybe thousands of galaxies, and even superclusters of several clusters.

Our galaxy, plus a couple of dozen others, form the Local Group of galaxies. This in turn is part of the Virgo Cluster. Such clusters are always dominated by giant elliptical galaxies, in each of which there are perhaps 500 billion stars – five times as many stars as in our galaxy.

Norma arm

Crux-Centaurus arm

Scutum arm

Sagittarius arm

Sun

Orion arm

Outer arm

The Sun is in the Orion Arm, *two-thirds of the way between the center and the edge of the galaxy, slightly to one side of the central plane.*

If the galaxy is looked at edge on, there is a central bulge of old stars. If the spiral arms were removed, this central region would look like an elliptical galaxy. The central region of the galaxy is strongly yellow, characteristic of the old stars found there. The spiral arms are dominated by bluish supergiant stars, *together with lines of gas clouds glowing pink. The spiral arms are not true structures – they are really density waves that flow around the galaxy, just as bunched-up traffic on a freeway can move along slowly, even though the vehicles are going at different speeds.*

*Our view of the Milky Way, looking toward the center (**right**), gives only a hint of the bulge, because nearby dust clouds hide the true center from us.*

How stars are born

New stars have been forming ever since the universe began in the Big Bang. And some new stars develop families of planets.

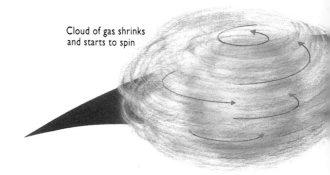

Cloud of gas shrinks and starts to spin

In space there are clouds that can make sunshine. Stars are, in fact, born from clouds of gas and dust scattered throughout the universe. But the material in these clouds, or nebulae, is extremely thin. There are just 16 million particles in every cubic inch of these space clouds, 100 billion times fewer than in ordinary air. Every now and then, parts of a cloud begin to clump together, squeezed perhaps by the explosion of a dying star or the pull of nearby stars. Then, over millions of years, the combined gravity of the particles pulls them together into hundreds of dark blobs, each the seed of a new star.

As these blobs shrink under the pull of their own gravity, they heat up – simply because of the movement of the particles speeding together. Eventually, the material around the center gets thick and opaque, and heat cannot escape. In a year or two, temperatures at the core soar to 18 million °F (10 million °C) and the new star forms, whirling around and bulging outward around its middle.

The core is now so hot that the nuclei of hydrogen begin to fuse, just as in a nuclear bomb, releasing huge amounts of energy. As the veils of opaque material are blown away to reveal this nuclear furnace, the star begins to shine. Meanwhile, any disk of material left over from the star's formation quickly begins to cool and condense into fragments. These in turn gather, or accrete, into lumps which may eventually form planets.

The cloud from which a star forms probably rotates only a little. But as it shrinks the spin increases, just as spinning skaters speed up by bringing in their arms. By the time the protostar has formed, the radiation can escape most easily at the flattened poles, creating twin outflow jets. Eventually jets of gas blast their way through the dark veil of clouds covering the star. If the star is less than a tenth the size of the Sun, the core temperature may not be high enough to trigger nuclear fusion, and it stays a brown dwarf.

When a star is big enough, fusion starts, and the star begins to radiate vast amounts of energy. Meanwhile, it is thought that around some stars spare gas and dust form into a disk in which material collects into chunks called planetoids. These collide with each other, and the chunks get larger, forming planets. Outer planets are far enough away from the new star to keep their light elements such as hydrogen and helium, but in the fierce glare close to a star, the planets have these elements stripped away, leaving them rocky and small.

Solar system forms

Outer planets retain light gases

Light gases driven off inner planets

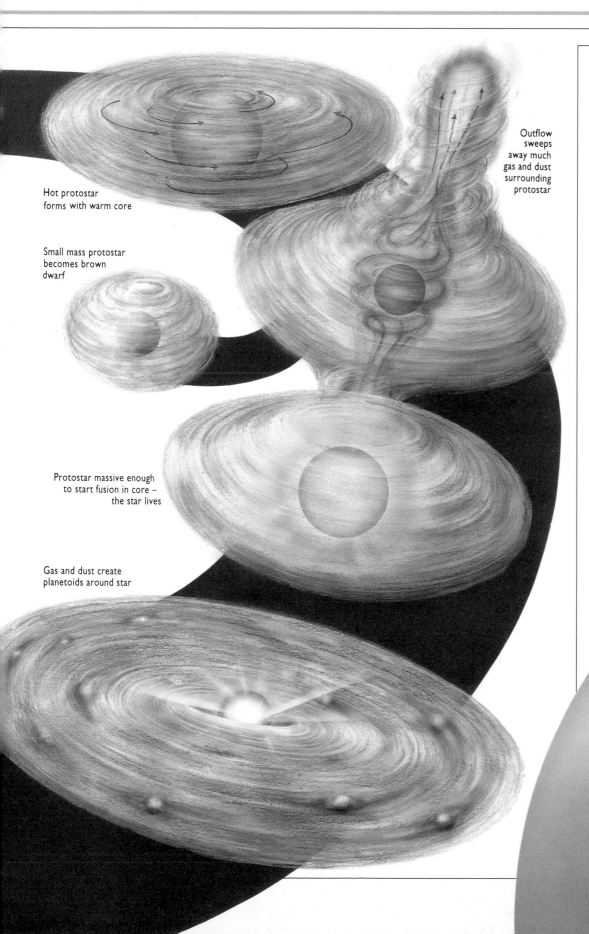

Hot protostar
forms with warm core

Small mass protostar
becomes brown
dwarf

Outflow
sweeps
away much
gas and dust
surrounding
protostar

Protostar massive enough
to start fusion in core –
the star lives

Gas and dust create
planetoids around star

SIZE AND COLOR

Stars can be one of a variety of colors. A star's color depends on its surface temperature, just as heated metals go from red to yellow to white and blue as they get hotter. Stars range from below 6,300°F (3,500°C) – red – to 45,000°F (25,000°C) – blue.

When they are born, most stars are close to the size of our Sun. The biggest blue stars are only about 20 times as wide as a Sun-sized star. Stars in this range are called main-sequence stars, and there is a direct link between size and color. The larger the star, the hotter and bluer it is.

With relatively newly born, medium-sized stars, there is also a simple link between color and brightness. The whiter and hotter a star, the more brightly it glows. The redder and cooler it is, the more dimly it glows.

See also

SPACE
▶ Our local
star
22/23

▶ Systems
of stars
34/35

ENERGY
▶ The energy
spectrum
70/71

▶ A matter
of degree
80/81

▶ Falling for
gravity
90/91

**ATOMS AND
MATTER**
▶ Energy
from atoms
142/143

Dim red star
Surface temperature
below 6,300°F (3,500°C)

Sun-size yellow star
Surface temperature
10,500°F (5,800°C)

Hot blue star
Surface temperature
20,000–45,000°F (11,000–25,000°C)

2
Red giant

The dying light

Death eventually comes to every star, as the supply of nuclear fuel that kept it going is burned out.

Hydrogen burns
steadily in core

1
Sun-size
star

Core shrinks and
outer layers swell

Helium-
burning
core

For some stars, death comes quickly. While the sedate Sun may take 10 billion years to use up its supply of nuclear fuel, the brightest, most massive stars can squander their supplies in just a few million years. At the other end of the scale, miserly red dwarfs will still be glowing feebly after 200 billion years.

The deaths of stars match their lives. Massive stars end in spectacular explosive collapse as supernovae, while a smaller star such as the Sun will, in about five billion years, transform itself into a swollen red giant before collapsing to end its days as a white dwarf. Red dwarfs may simply fade away at the end of their lives.

All these changes are governed by what happens inside the star when the hydrogen and helium fuels run low. When the reactions that power the star can no longer sustain it against the pull of gravity, drastic changes in its structure occur. In very large stars, the pressure and temperature in the core can be so great that even relatively heavy elements can become nuclear fuels, fusing together to form iron. Eventually, however, there is a catastrophic collapse, as the star shrinks to almost nothing in less than a second, then rebounds in a massive supernova explosion.

When a star the size of the Sun uses up its fuel, it begins to rearrange itself by blowing off giant shells of matter, which billow off into space as planetary nebulae, leaving only the hot core. In the far future when our Sun becomes a red giant, Earth will be incinerated.

*Stars such as the Sun spend 10 billion years shining steadily (**1**). When their normal fuel runs low, the outer layers puff out to create a huge red giant star (**2**), and the core shrinks so that its material condenses into dense degenerate matter in which the atomic particles are crushed together. Eventually, the star sheds more material to relieve the pressure of its outer layers. It throws off shells every few thousand years, losing about 10 percent of its mass each time, until its core is about half the size. The shells become a planetary nebula (**3**). The hot ember is an Earth-sized white dwarf star of degenerate matter (**4**) that shines feebly as it cools down.*

Star collapses,
contracts, then
explodes

4
Neutron star

4
Black hole

3
Supernova

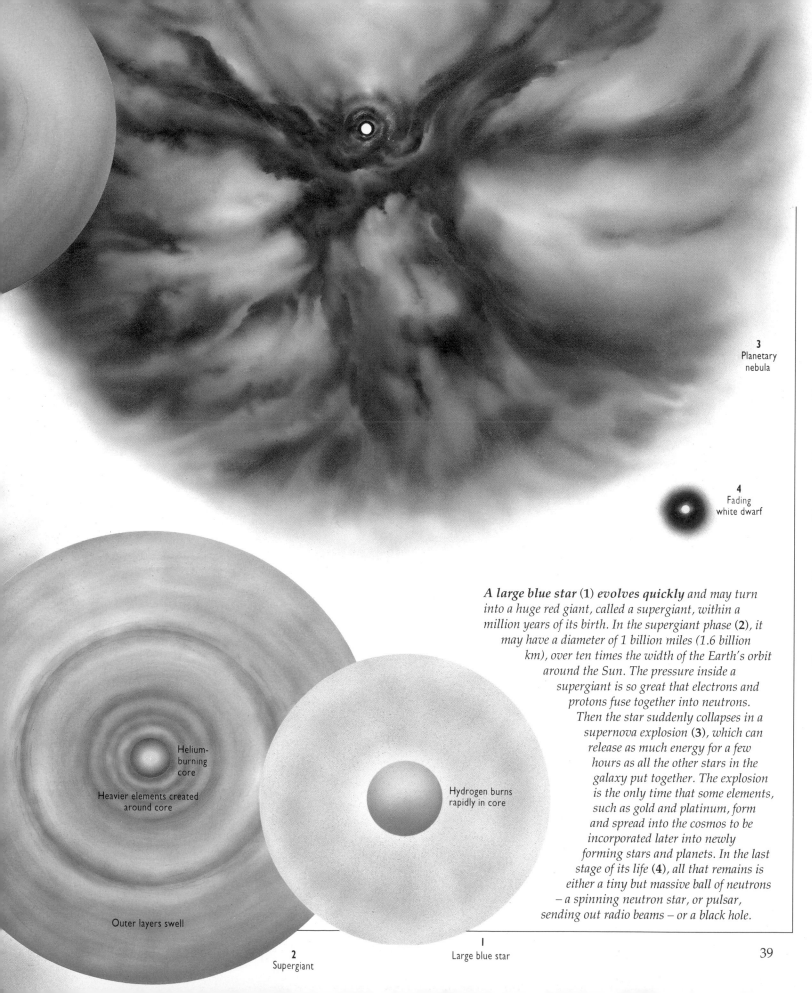

3
Planetary
nebula

4
Fading
white dwarf

A large blue star (**1**) *evolves quickly and may turn
into a huge red giant, called a supergiant, within a
million years of its birth. In the supergiant phase (**2**), it
may have a diameter of 1 billion miles (1.6 billion
km), over ten times the width of the Earth's orbit
around the Sun. The pressure inside a
supergiant is so great that electrons and
protons fuse together into neutrons.
Then the star suddenly collapses in a
supernova explosion (**3**), which can
release as much energy for a few
hours as all the other stars in the
galaxy put together. The explosion
is the only time that some elements,
such as gold and platinum, form
and spread into the cosmos to be
incorporated later into newly
forming stars and planets. In the last
stage of its life (**4**), all that remains is
either a tiny but massive ball of neutrons
– a spinning neutron star, or pulsar,
sending out radio beams – or a black hole.*

Helium-
burning
core

Heavier elements created
around core

Outer layers swell

Hydrogen burns
rapidly in core

2
Supergiant

I
Large blue star

Black holes

The laws of physics predict there are regions of space where gravity is so strong that nothing, not even light, can escape. But no one can directly observe them.

Astrophysicists hypothesize that in a black hole matter is squeezed so powerfully that its gravity becomes irresistible. A black hole the same size as the Sun would contain more than a billion times as much matter as the Sun itself. The gravitational pull of a black hole this size would be so immense that anything within range would be sucked inexorably into its black heart. Even light could not escape. This is why it is called a black hole.

Some astrophysicists believe that the biggest supernovae may end up as black holes. The pressure of so much material in a massive star may be so great that when the star dies and collapses, the collapse continues, and the star gets even denser and so shrinks faster and faster.

But the most massive black holes may be associated not with the death of stars, but with mysterious objects called quasars – the most intense energy sources in the universe, as bright as 100 galaxies, yet no bigger than our solar system. Quasars are billions of light years away, yet emit strong radio signals that are easy to pick up. Observe an object billions of light years away, and you see it as it was billions of years ago, because the electromagnetic radiation from it has taken that length of time to reach you. Quasars thus show what was going on in the universe billions of years ago when it was very young.

The theory is that, as galaxies formed in the early universe, powerful gravitational attraction pulled huge amounts of gas together at their center to create a black hole. This sucked in not only gas but any stars that strayed close, shredding them to gas in the process. As the gas swirled into the hole in a huge vortex, it got hotter and hotter and glowed more and more brightly. Just before its plunge into oblivion, it emitted a burst of radiation, including massive flashes of X-ray and ultraviolet radiation. A quasar is thought to be this final burst.

Despite being used as explanations for things such as quasars, there is a lack of direct evidence for black holes. This is not surprising since, by definition, they cannot be seen. But their effects on surrounding matter can be detected, and astrophysicists are confident that black holes exist.

The giant elliptical galaxy M87 has long been known to possess a jet of light extending over 5,000 light-years from the core of the galaxy. This Hubble Space Telescope picture reveals for the first time a bright spot at the galaxy's heart. It is believed to be not starlight, but a hotspot caused by gas falling into a black hole with a mass 2.6 billion times that of the Sun.

Scientists believe that the jets of light are actually intense electric streams generated by the swirling gases. It is these huge electric fountains that are thought to create the strong radio signals picked up from quasars, short for "quasi-stellar radio sources." With radio telescopes, similar large jets can be seen shooting out of many quasars.

Even when there is no sign of a bright quasar, such jets are often evident in "radio galaxies," galaxies that emit a stream of powerful radio signals. Here it is thought that the bright disk of the quasar around the central black hole may have faded away to nothing.

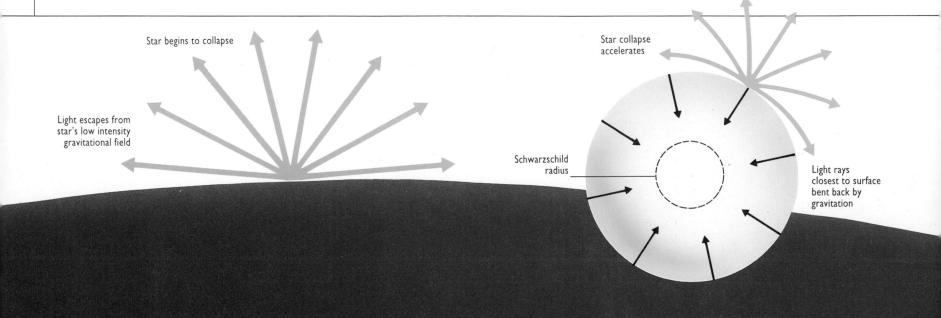

Star begins to collapse

Star collapse accelerates

Light escapes from star's low intensity gravitational field

Schwarzschild radius

Light rays closest to surface bent back by gravitation

LOST LIGHT AND TIME

The gravitational field of a supergiant is not strong enough to bend back the light it radiates in any significant way. When the star dies and its core collapses, the gravity field close to it becomes proportionally stronger the smaller the star gets. Light continues to be radiated by the star until it collapses to below a certain size, called the Schwarzschild radius. Once smaller than this, the gravity is so strong that all light is bent back, and a black hole forms. The Schwarzschild radius then becomes the hole's event horizon – the region around the collapsed star in which events inside the star cannot be detected.

As the collapse continues, the whole star is compressed into a singularity – a single point – and the laws of matter and time, as we know them, do not apply.

Just before star reaches
Schwarzschild radius, only
vertical rays can escape

No light can escape
once star reaches
Schwarzschild radius

Entire mass of
star collapses
into black hole

Singularity

Event horizon

Creation of the cosmos

The universe is thought to have begun in a huge explosion whose echoes still reverberate through space.

Most astrophysicists accept the theory that the universe began with the biggest explosion of all time, called the Big Bang. One moment there was just an unimaginably small and hot ball – smaller than an atom. A moment later the universe exploded into existence, with a bang so big that material is still hurtling away from it in all directions at astonishing speeds.

There is plenty of evidence that the universe is expanding. Some of the best comes from observing the light from distant objects. Light is shifted toward the red end of the spectrum when any object being watched is moving away from the observer. The faster an object is moving away, the farther its light will be shifted to the red. The red shift detected in the light from every distant galaxy, cluster of galaxies, or quasar shows that they are moving away from us and also from one another. And the more distant an object is, the more its light is red-shifted, and thus the faster away it is moving. Run this expansion backward and at some time, between 14 and 20 billion years ago, everything comes together. A clinching piece of evidence for this scenario is the faint glow still detectable wherever we look in the sky – what is called the microwave background radiation.

There have been other explanations for the expansion. One popular idea was the steady-state theory – the idea that the expansion is making way for new matter being created all the time. But this theory cannot explain the microwave background.

The history of the universe since the Big Bang has been calculated in some detail. The universe began with just a minute, incredibly hot ball. But within a tiny fraction of a second after the Big Bang, it was swelling by an astonishing hundred billion billion billion billion times – a process called inflation. While it rapidly mushroomed out, the infant universe began to cool, and matter and the basic forces such as electricity and magnetism were created.

At first, matter consisted only of basic particles called quarks, but within about a billionth of a second, they began to join up in the familiar combinations that we call protons, neutrons, and electrons. Then neutrons and protons combined to form the simplest atoms, first hydrogen and then helium, with three-quarters hydrogen and a quarter helium. The universe was filled with swirling clouds of gas. These gradually curdled into long thin strands which eventually clumped together to form galaxies, stars, and planets.

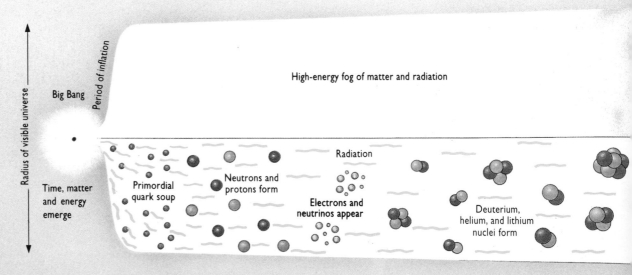

Radius of visible universe

Big Bang

Period of inflation

High-energy fog of matter and radiation

Time, matter and energy emerge

Primordial quark soup

Neutrons and protons form

Radiation

Electrons and neutrinos appear

Deuterium, helium, and lithium nuclei form

Time since Big Bang   0 seconds 10^{-43} 10^{-34} 10^{-10} 10^{-6} 10^{-4} 1 second   1 minute   1 week   10,000 years

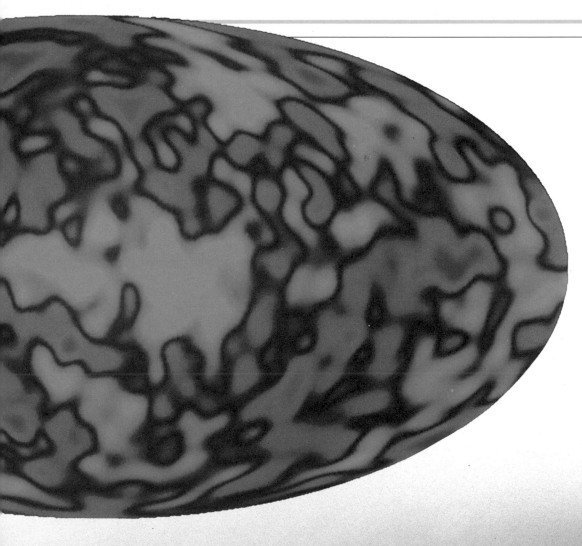

Galaxies and stars formed when gravity began to pull the swirling gas clouds of the primordial universe into clumps. Yet scientists puzzle over how this happened, for there is not enough matter for gravity to have pulled it together so quickly.

Indeed, the galaxies we see today could only have formed this way if there was perhaps 100 times as much matter in the universe as there seems to be. This is why some people believe the universe is made mostly of cold dark matter which we cannot see.

But even the idea of dark matter does not explain why the gases clumped together in the first place. If the Big Bang theory is correct, the universe must have been slightly lumpy very early on.

This is why this map of the whole sky made using images from the COBE satellite in 1992 was hailed as a breakthrough. It shows ripples in the microwave background. This means that 300,000 years after the Big Bang, the universe was not totally smooth and primitive structures were beginning to form.

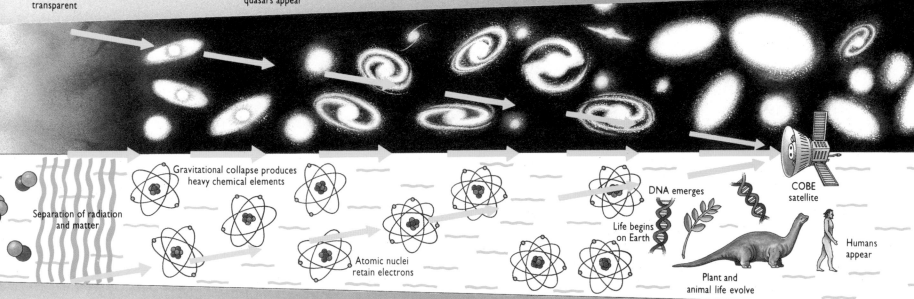

Universe becomes transparent

First galaxies and quasars appear

Modern universe

Separation of radiation and matter

Gravitational collapse produces heavy chemical elements

Atomic nuclei retain electrons

DNA emerges

Life begins on Earth

Plant and animal life evolve

COBE satellite

Humans appear

300,000 years   1 million years   1 billion years   4 billion years   12 billion years   15 billion years

The unfolding universe

Will the universe go on forever, or will it come to an end as spectacular as the way it began?

Whatever happens to the universe as a whole, there is no doubt that our own Sun will eventually die. At the moment the Sun is about five billion years old and roughly halfway through its life span. Over the next five billion years, it will gradually get bigger and hotter, causing unpredictable changes to the Earth's climate. Then, it will begin to die, leaving the Earth a cold globe orbiting the tiny spark of a white dwarf.

While stars like the Sun die, new stars are born, but the process cannot go on forever. After 100 billion billion years, there will be no gas left to form new stars, and the last remaining red dwarfs will be cinders. Every star in the universe will eventually dwindle to a cold dark lump.

Moreover, although the universe is still expanding now, there is no certainty that it will go on doing so indefinitely. Some scientists believe gravity is already beginning to put a brake on its expansion. Many believe that if enough invisible dark matter does in fact exist in the universe, its combined gravity may be enough to stop expansion altogether. If this happens, gravity may then pull all the galaxies together again in a massive Big Crunch, like the Big Bang in reverse. No one knows quite what would follow a Big Crunch, but some maintain that if all the energy and matter in the universe were crushed into a tiny ball in a Big Crunch, it would at once bounce back in a second Big Bang, and the universe could go on alternately expanding and contracting forever.

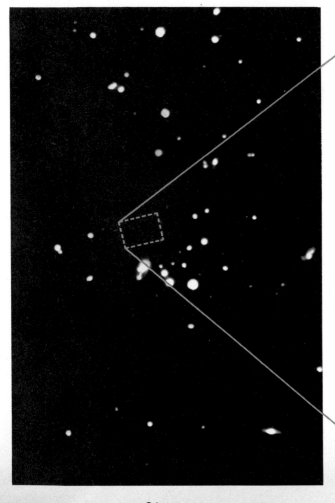

Universe stops expanding and starts to contract   Clusters of galaxies begin to fall together

Galaxies merge as temperature of universe rises

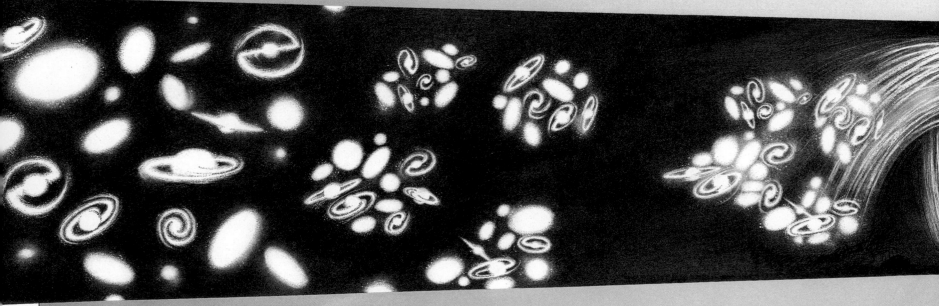

50 billion years   1 billion years   100 million years

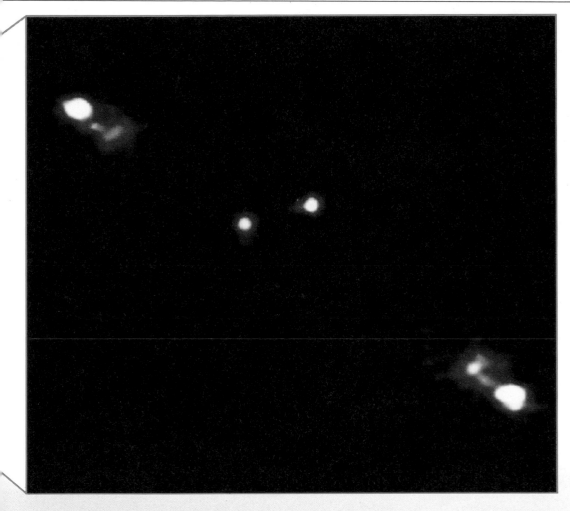

A crucial factor in the eventual fate of the universe is the amount of matter it contains. If it contains no more than we can see, the chances are that gravity will never be enough to stop the galaxies from hurtling apart. This means the universe will go on expanding forever, an idea called the "open universe" concept. If, however, it contains all the invisible dark matter that some suspect, the expansion will almost certainly be brought to a halt, and the universe will then contract in a Big Crunch.

   Significant evidence of the existence of dark matter was provided by this picture from the Hubble Space Telescope. The pattern it shows is not an optical illusion, but a gravitational one. In the center is a faint galaxy some 400 million light years away. Behind it lies a quasar 20 times more distant. The galaxy's gravity has bent the quasar's light into four images – called an Einstein Cross. Hubble Space Telescope observations of the numbers of such crosses help to reveal dark matter and to evaluate rival cosmological theories.

Even the most distant galaxies and quasars we can see show that the universe is expanding. But these objects may, in the far future, begin to slow their expansion and stop moving apart, as if linked to one another by elastic. This elastic would be the dark matter invisible to us: if this is widespread, it would mean that the universe is much more massive than meets the eye. It is revealed by its effects on the movements of galaxies, which suggest that there may be 100 times more dark matter than the material we can observe. If so, the elastic may be strong enough to pull the galaxies back together again. At the moment, it seems the universe is hovering between infinite expansion and eventual contraction.

Stars move nearly as fast
as speed of light around
hypergalactic center

Space becomes so hot that
all stars are destroyed

Black holes grow as
temperature rises
even higher

Universe collapses
to giant fireball,
like Big Bang

Big Crunch

Expansion
begins again?

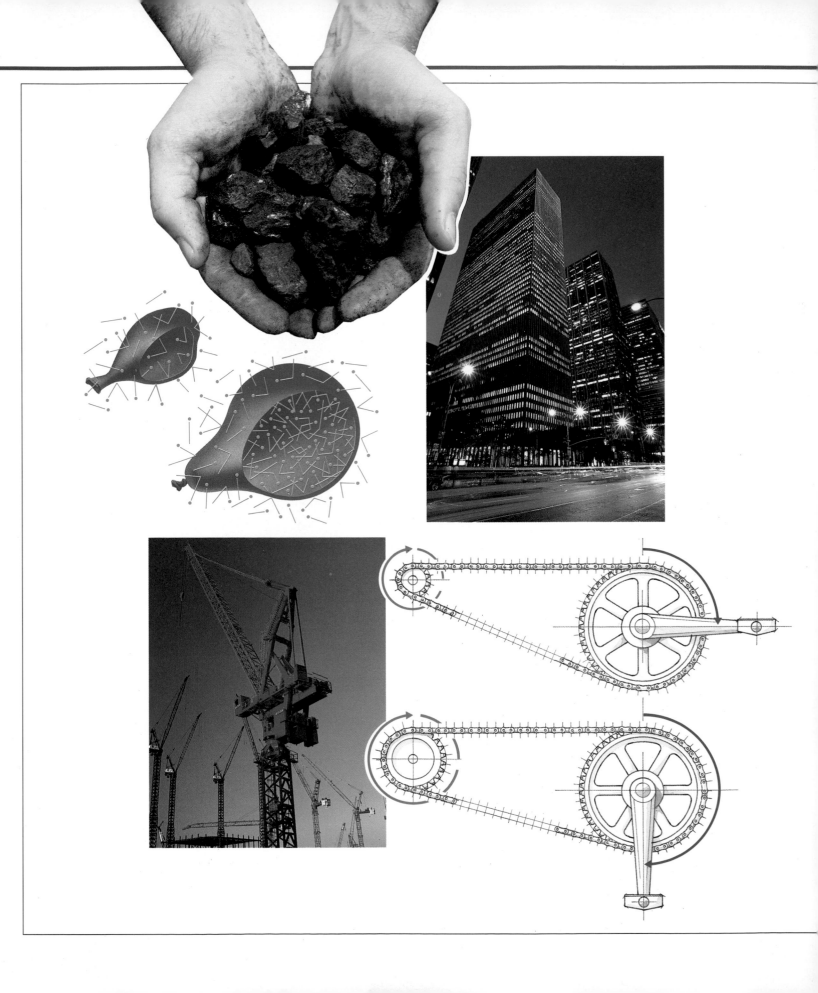

Energy

Energy makes things happen. A battery has energy: it can light a flashlight bulb. Boiling water has energy: it can drive a steam engine. A ball in flight has energy: it can smash a window. Light has energy: it can make a plant grow.

Energy is also conserved. Through all the processes and changes that it brings about, it survives undiminished, in one form or another, able to bring about new changes. This is why scientists are so persistent in tracing the transformations of energy in our world from the heart of the atom to the laws governing the lives and deaths of galaxies.

The study of energy and its various manifestations and forms is at the heart of understanding the science of the world in which we live. The idea of energy used only to be applied to mechanical processes – lifting weights and turning wheels. Then it was found that mechanical energy could be converted into heat and back. Soon electrical, magnetic, nuclear, and other forms of energy were added to the types of energy which exist.

Left clockwise from top: fossil fuel; using electricity; bicycle and force transmission; cranes and levers; pressure and gases.
This page (top): *Light energy and refraction;* **(left)** *force of gravity.*

Making things move

Everywhere on Earth, the vital links between energy, force, and work allow objects to move from place to place.

Everything in the world is constantly pushed and pulled by unseen forces. A book lying on a table might look as if nothing is happening to it, but it is, in fact, held there by two balanced forces. The Earth's gravity is responsible for the first force by acting on the book to give it weight – a force acting downward. The second force comes from the table which opposes this downward force with an equal, upward force. But if the book is pushed over the edge of the table, there is nothing to resist the downward force and the book falls to the floor. To put the book back on the table, someone picking the book up has to exert a force greater than the weight of the book to move it off the floor. In exerting force and moving the book up, that person does work.

Whenever a force moves an object, work is done, and energy changes are involved. Lifting the book to the table gives the book potential (stored) energy. If an object has potential energy, then it can do work and produce other forms of energy. The book's gain in potential energy comes from the person who loses (uses up) physical energy by working against gravity to lift the book.

The book's potential energy can be released and turned into another form of energy called kinetic energy (energy of motion) if the book falls back to the floor. It is the work done by the downward force – the weight – of the book that creates the kinetic energy. When the book hits the floor, its kinetic energy may be enough to dent the floor or deform the book, as well as to make sound energy, which is the noise the book makes as it contacts the floor.

Several forces often combine to create a single force. For instance, the thrust of an aircraft's engines and the sideways push from a crosswind combine and the resultant force makes the aircraft drift slightly to the side as it flies forward.

Small forces can seem to work miracles. For instance, a plant growing in a minute crack in a rock can eventually split the rock apart. The force needed to split a rock might seem beyond the power of a mere plant. A person would have to use a massive hammer and chisel to achieve the same result.

A plant's secret is that it exerts force over a long time. At first, the work done by a plant is provided by energy that came from its seed's food stores. Later, it uses energy from the Sun to power its growth. As the roots expand and grow, they can slowly force rock apart.

Steep slope

Weight due to gravity
Downslope force
Force acting on surface
Surface reaction force

Gentle slope

Gravity acts to make the skier hurtle down the slope. The skier's weight, a force acting straight down, is given by gravity. On a slope the force is split into two parts: one at a right angle to the slope, the other parallel to it.

The force at a right angle is the part of the skier's weight the snow supports. Because the skis slip over the snow, the parallel force, the part of the skier's weight acting down the slope, is not opposed, so the skier accelerates downhill. The steeper the slope, the greater the parallel component, so the faster the skier accelerates.

As the skier accelerates, work is done. When work is done, energy is transferred, and in this case the skier gains kinetic energy (energy of motion). The skier at the top of the slope has potential energy.

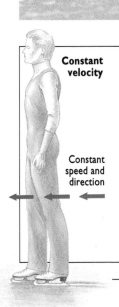

Constant velocity

Constant speed and direction

See also

▶ **ENERGY**
Working under pressure
56/57

▶ Energy's endless flow
58/59

▶ Falling for gravity
90/91

SPACE
▶ The giver of life
24/25

LIFE
▶ Light for life
154/155

HOW FORCES MOVE OBJECTS

Three laws sum up how force moves objects. First, any object stays at rest or continues moving in a straight line at a constant speed until some force acts on it. So if a skater (**left**) does not try to go faster or slower, he just carries straight on at the same speed. Second, a force changes an object's velocity. So if a skater is towed (**right**), the force accelerates her, changing her velocity. Third, action and reaction are equal and opposite. So when a skater pushes another (**far right**), the forces are the same, but the lighter skater is accelerated more.

Changing velocity

Force applied

Velocity increases

Action and reaction

Equal and opposite forces

Lighter skater, faster speed

Heavier skater, slower speed

49

Simple machines

In almost every situation, there is a machine to help make work easier. The most common machines are some of the simplest.

There is a huge variety of machines. These mechanical devices that make possible all kinds of human activities include machines that convert energy from one form to another, as in the gasoline engine which converts the stored chemical energy of gas into energy of movement. The word machine is also used for things that merely process information, such as television sets and computers.

A simple machine is something very specific. It is a device that can carry force from one place to another, change its direction, or change its strength. The crowbar used to lift a floorboard is a simple machine. Another simple machine is the screw: its action is seen clearly in a car jack, which converts the force of a person's hand turning a handle into a force large enough to lift a car. Similarly, the wood screw is driven into lumber by the weak force of fingers turning a screwdriver.

Simple machines that multiply force have one thing in common: if they are to work, the force has to be moved farther than the load is moved. The hand pressing on the crowbar moves about 10 times as far as the corner of the board that is raised. The hand on a jack handle moves a long way (because it makes many turns) to raise the car a short way. This is how all simple machines work. They convert a small force, or effort, exerted over a long distance, into a larger force, or load, applied over a shorter distance.

In fact, the work done by the operator of a machine (effort times distance moved by effort) equals the work done on the object moved (load times distance moved by load). A heavy load can be moved because the weak effort is spread out over a longer distance. So the secret of building a force-multiplying machine is to make sure the effort moves a greater distance than the load. For instance, a pulley is designed so that when a great length of rope is pulled, the load moves only a short way. But machines do not only increase force. Machines like tweezers and barbecue tongs reduce force because the ends of the tongs move farther than the user's fingertips and so exert a smaller force.

When work is done, energy is transferred or converted. So when a load is raised against the force of gravity, it gains potential energy. When a screw is driven into lumber, some of the work done is converted into heat.

Large force over short distance

See also

ENERGY
▶ Making things move
48/49

▶ Machines in motion
52/53

▶ Energy's endless flow
58/59

▶ Falling for gravity
90/91

BRAINS AND COMPUTERS
▶ Calculating machines
216/217

Levers can help pull a cork from a bottle. With bare hands it would be impossible to get the cork out. For a start, the cork is inaccessible; and even if the cork was protruding from the bottle a little, it would be hard to get a good grip on it using just fingers. The answer to getting at the cork is to use a screw which can be driven into the cork without too much effort. A large movement of the fingers on the end of the screw handle is translated into a short movement into the cork so the force of the fingers is multiplied. The next problem is to remove the cork. With an ordinary corkscrew (**below**), the cork has to be pulled out using brute force. The force is applied directly and has to be more than the force caused by the friction as the neck of the bottle squeezes the cork. Much less force is needed if a corkscrew with arms is used (**left**). The arms act as the long arm of a lever, so when the hand moves, applying a moderate force to the end of the arms, this is converted into a much stronger force that easily pulls the cork from the bottle.

KEEPING A BALANCE

The seesaw is a good example of how a lever works. Two people of unequal weight cannot balance if they sit the same distance from the fulcrum of the seesaw. For a system to be in balance, the force on one side times its distance from the fulcrum must be the same as the force on the other side times its distance from the fulcrum. As the heavier person moves toward the center, there comes a point when the small force of the small person a long way from the center matches the larger force of the heavier person, now closer to the center, and the seesaw balances.

Out of balance

Equal distance from fulcrum

Small force

Fulcrum

Large force

In balance

Short distance

Long distance

Fulcrum

Large force

Small force

Small effort over long distance

Large effort over short distance

Large force over short distance

Large force over short distance

Small effort over long distance

Fulcrum

Resistance of shell

*A **nutcracker works** using a pair of levers. The load (the force of the shell of the nut resisting the cracker) is between the effort (the force of the fingers squeezing the tips of the handles) and the fulcrum (the hinge of the nutcracker).*

The comparatively weak force applied by the fingers moves through a greater distance than the comparatively strong force delivered *at the inner ends of the handles. The force of the fingers is enough, when multiplied by the nutcracker, to crack open the shell and release the nut hiding inside.*

Machines in motion

Complex machines are so much part of everyday life that it is easy to forget that they actually operate on very simple principles.

With the help of simple machines such as the lever and the pulley, which increase or redirect the force of our muscles, we can perform feats that would be impossible with our bare hands. In everyday life we make use of a huge variety of machines – from giant cranes to power generators and even the humble bicycle. But all machines, no matter how complex, use the basic principles of simple machines, especially the lever.

Levers act not only in straight lines, but can also multiply force as they rotate. For instance, in a wind turbine, the rotor blades harness wind power and transfer it to a central axle (shaft). Although the axle rotates through the same angle as the end of the rotor blade, it moves a much smaller distance. This means that the relatively small force from the wind as it pushes the end of the blade through a large distance is translated into a much stronger force at the axle. The power in the axle that came from a puff of wind can then be used to turn a generator and make electricity.

Gears are used in a huge number of machines from cars to computer printers; they are toothed wheels which, when meshed together, transfer and modify force. When a large gear rotates, the strong force at its axle is transferred and reduced to a much smaller force in the teeth at its rim. When this force drives a smaller gear, the smaller gear ends up with less force at its axle than the large gear because the force at its rim is multiplied by a shorter distance. Using gears of different sizes, forces at axles can be increased or reduced as needed.

Cranes lift heavy weights from place to place. The crane's long arm, or jib, is a huge lever which has its pivot, or fulcrum, directly over the top of the crane's supporting tower. The crane has a square-shaped counterweight just to one side of the fulcrum to balance the force from the load at the other end of the jib.

When the jib is lowered to pick up a weight some distance from the tower, the force exerted by the object being picked up increases sharply. This would buckle the jib if extra support were not given by the strong cables that stretch between the crane top, the counterweight and the end of the jib.

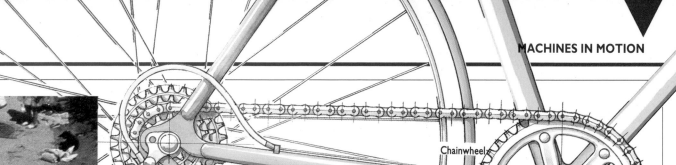

Chainwheel

Rear sprockets

Derailleur

Crank

Pedal

Chain

Rear wheel

See also

▶ **ENERGY**
Making
things move
48/49

▶ Simple
machines
50/51

▶ Energy
without limit
62/63

▶ Generating
electricity
68/69

▷ **SPACE**
The giver
of life
24/25

The pedals of a bicycle turn the cranks which act as levers to rotate the chainwheel (**above**). The chain connects the chainwheel's sprocket to sprockets of various sizes driving the rear wheel. Sprockets act like gearwheels, but instead of meshing directly they are linked by a chain. Thus the chain transmits force from the feet on the pedals into force that drives the rear wheel. The derailleur mechanism moves the chain between the rear sprockets and is used by the cyclist to select the different sprockets, or gears.

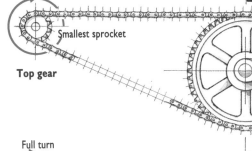

Full turn

Smallest sprocket

Quarter turn

Top gear

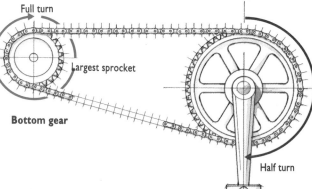

Full turn

Largest sprocket

Bottom gear

Half turn

Gears let a cyclist pedal with about the same effort on a level road or uphill slope. Most chainwheels have 48 teeth, so one complete turn moves the chain on by 48 links. On level ground, the cyclist selects a small rear sprocket (top gear) with just 12 teeth. This means it takes only a quarter turn of the chainwheel to rotate the rear sprocket, and the rear wheel, once.

Going uphill, the cyclist selects a larger rear sprocket (bottom gear) with 24 teeth. It now takes half a turn of the chainwheel to rotate the rear sprocket, and the rear wheel, once. The force at the rim of the wheel is doubled in bottom gear, and the cyclist can tackle the hill, but goes only half the distance for the same work done in top gear.

The bicycle is one of the most efficient machines ever invented for transforming force into forward motion. These cyclists (**left**) competing in the Tour de France race keep their bicycles moving for hours at a time at speeds they could not possibly reach on foot.

What is pressure?

*Pressure makes water gush from hoses, keeps the bubbles in a soda pop,
and stretches the sides of a balloon when you blow it up.*

Walk on snow wearing shoes and you sink, because shoes concentrate the force of your weight on a small area. The force per unit area, or pressure, is high enough to penetrate the surface. But put on skis and they spread the same force over a larger area, so the pressure is less and you do not sink.

In solids like snow, pressure is absorbed locally because the atoms in solids are fixed in place. But alter the volume of a gas or apply force to a liquid, and the pressure changes throughout because the atoms (joined together as molecules) move freely.

In a gas or liquid, the impacts of molecules hitting a surface create a force and thus pressure on it. The force depends on the number of impacts, the mass of the molecules, and their speed. Compress a gas, and the same number of molecules occupies a smaller space, so there are more impacts and the pressure rises. Heat a gas without changing its volume, and the molecules move faster, so each impact is harder and pressure rises. With pressure changes come energy transformations. For instance, the work done pumping up a tire turns to heat energy and potential energy.

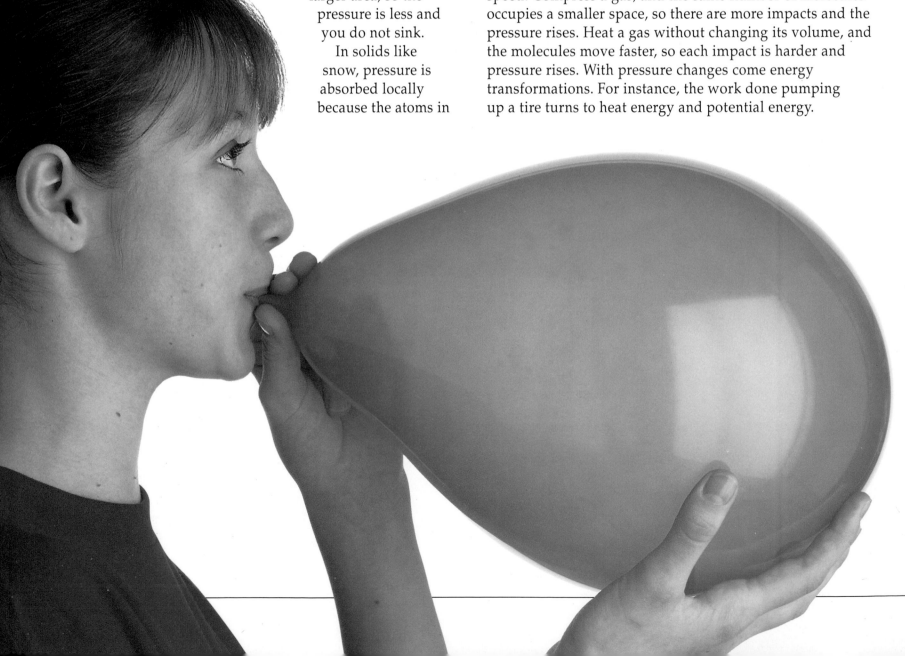

Before a balloon is inflated, the pressures inside and outside it are equal so the skin of the balloon is flaccid. Air pressure comes from the force of the countless gas molecules bombarding a surface. Inside an uninflated balloon, there are the same number of molecules per unit volume as outside, so the number of impacts is the same and thus the pressure is equal.

When a balloon is blown up, a huge number of extra molecules are forced into the balloon's confined space. This causes lots of extra collisions, both between molecules and against the inside of the balloon. These collisions push on the balloon wall, raising pressure. When the force on the inside is greater than both the elastic tension of the rubber and the air pressure outside, the balloon expands.

Deflated balloon

Equal pressure from air molecules inside and outside balloon

Inflated balloon

More air molecules bombard inside of balloon raising pressure and stretching elastic wall

Firefighters use high-pressure jets of water to deliver water right to the heart of a fire. Hoses are attached to the main water supply, which is under pressure because the supply pipe is connected to a reservoir higher than the pipe.

Liquids show little change in volume as pressure alters. A pressure buildup can only be released by the liquid flowing to a place of lower pressure, for example, by gushing from a faucet which is turned on. The greater the difference in pressure between two places, the faster the fluid flows. Motors pump the water for fire hoses to give an even more powerful jet when the nozzle is opened.

BUBBLING UP

Putting the bubbles into soda pop is possible because gases dissolve in liquids under pressure. Carbon dioxide, the gas in such bubbles, is injected into a drink at high pressure and stays dissolved. When the bottle is opened, the pressure is released; the gas can no longer remain dissolved, so it bubbles out of the liquid.

Exactly the same process can lead to the bends, a serious problem for deep-sea divers. Underwater, divers breathe air stored under pressure in tanks. As this air enters the diver's lungs, the pressure makes the nitrogen in the air dissolve in the blood. If the diver surfaces too quickly, the rapid release of pressure forms dangerous nitrogen bubbles in the blood.

Working under pressure

Pressure is put to work in countless ways. It is used in the brakes of cars and in fire extinguishers, and it keeps aircraft flying in the sky.

Many everyday mechanisms make use of pressure in liquids and gases. A machine that uses liquid pressure is called an hydraulic machine. Car brakes are hydraulic because they make use of the fact that pressure applied to any part of a liquid in a sealed container, or system of pipes and cylinders, is felt all through the liquid and is more or less equal everywhere. The brake pedal is one end of a lever which presses a piston in a cylinder. So when the driver presses a foot on the pedal, the force of the piston creates pressure which is felt throughout the brake fluid. At the brakes, the force presses on another piston, which pushes the brake pads onto the brake disk. Friction between the pads and the disk slows the car down.

A device using air pressure is called a pneumatic machine. Typical is a vacuum cleaner which, like many other pneumatic machines, uses the pressure of the atmosphere to help it do its work. A fan, turned by the cleaner's electric motor, lowers air pressure on one side of itself and raises it on the other. On the side where the fan lowers pressure, the pressure of the atmosphere outside makes air rush into the machine through the hose, dragging dust and dirt with it. On the side where the pressure is raised, the air, now containing dirt particles, is forced through the dust bag. The air can leave only through the tiny spaces between the fibers of the dust bag, so only the finest dust particles escape – most that do are trapped by an additional filter so that only clean air is sent back out. The name "vacuum cleaner" is not in fact correct as the air pressure is only slightly lowered. More accurate would be "partial vacuum cleaner."

*The shiny metal tubes on the excavator (**left**) are parts of the hydraulic rams which give the arm and bucket the power to dig. Each ram has a piston in a cylinder linked to a reservoir of high-pressure hydraulic fluid. The driver opens valves to let the fluid into one end of the cylinder. The piston then moves in or out with a huge force which powers levers to move the arm or bucket.*

*The pressure in a moving gas is lower than in surrounding still gas. This becomes obvious on blowing a gentle stream of air above a piece of paper held so it hangs down (**below**). On blowing, the air moves, the pressure above the paper goes down, and the atmospheric pressure below the paper pushes the paper up.*

Lower pressure

Normal pressure

56

See also

► **ENERGY**
Simple
machines
50/51

► What is
pressure?
54/55

► Energy's
endless flow
58/59

► **SPACE**
The giver
of life
24/25

► **ATOMS AND
MATTER**
Molecules
in motion
98/99

► Gases: atoms
set free
100/101

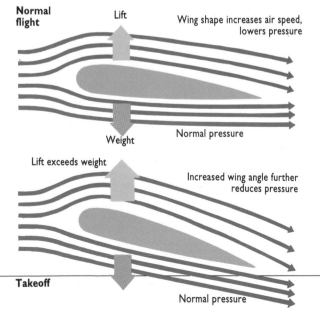

**Normal
flight**

Lift

Wing shape increases air speed,
lowers pressure

Weight

Normal pressure

Lift exceeds weight

Increased wing angle further
reduces pressure

Takeoff

Normal pressure

As in all planes, the Concorde has wings that use air pressure to hold it up. Wings have a curved top surface, which makes the air flow faster above the wing than below the wing. Pressure is lower in a moving, or faster-moving, gas, so air pressure falls above the wing. Low pressure above and normal pressure below cause an upward force on the wing, known as lift. Takeoff needs more lift, so the wing angle is steeper to speed the air above still more.

Many aircraft also have flaps on the edges of their wings which can be extended to give extra lift at a time when the plane is moving slowly through the air, for instance, during takeoff and landing.

Pilots have to be constantly aware of the weather, which is itself driven by pressure changes in the air. It is, after all, pressure differences in the air that cause the wind to blow. Pilots always receive up-to-date weather reports for the air they will fly through before they set off.

Measuring the pressure of the atmosphere helps forecasters predict the weather. Air pressure is mapped on charts by isobars, lines connecting places of equal pressure. Low pressure usually means bad weather and high pressure settled weather.

Energy's endless flow

Continually changing from one form to another, energy is unique – you can use it, but you can never lose it.

Wherever there is movement or change, there is a flow of energy. Energy comes in many forms. It is in the light that streams from the Sun and helps plants grow. It is in the heat that evaporates water from a glass and spreads it as invisible vapor through a room. It is in the electricity that drives a motor or heats up an oven. It is in the wind that fills the sails of a yacht and sends it skimming over the sea. In each of these examples of energy in action, the energy changes its form or is stored in a form different from the original.

Although there are many types of stored, or potential, energy, all of them can be converted into other sorts of energy. For example, many chemical substances contain chemical potential energy, which can be released in a reaction. When this happens rapidly – as in a firework – it can cause an explosion. When it happens slowly – as in a battery – it can drive an electric motor. In an explosion, the stored energy is turned into heat, light, and sound. In the battery the stored energy is turned into electrical energy, which itself can be used to power some device that changes the electrical energy into another form of energy, maybe the glow of a flashlight bulb, or the sound from a personal stereo.

When energy changes from one type to another, the total amount of energy does not alter. Energy is neither lost nor gained, just transformed or stored. The fact that energy is neither created nor destroyed sums up one very important law, the law of conservation of energy. In fact, use a certain amount of physical energy to hoist a weight to the top of a tree, and the weight gives back the identical amount of energy when it comes down to the starting point. Switch on an electric light bulb, and the heat and light energy generated from the electricity would, if they could be converted back again, produce the same electrical energy as you started with.

*It takes an enormous amount of energy to power the world's cars (**right**). But a journey in a car neither creates nor destroys energy, it transforms it (**below**).*

Gasoline, containing chemical energy, is burned in the car engine to produce heat energy, which expands the gases in the cylinder, forcing the piston down. This motion is converted to rotary motion and transferred to the wheels. These move the car, giving it kinetic energy (energy of motion).

As the engine works, some heat is lost in the exhaust gases, and a small amount of energy is lost as vibration energy, or sound, created by the engine. When the car is moving, friction comes into play as the car rushes through the air and as its wheels make contact with the road. As a result, heat is lost.

Energy is also converted to electrical energy to power the car's electrical components, such as the headlights and brake lights. When the car is slowed, the brake pads and disks turn kinetic energy to heat and sound.

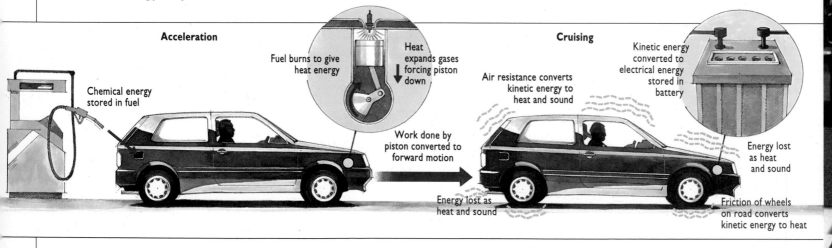

Acceleration

Chemical energy stored in fuel

Fuel burns to give heat energy

Heat expands gases forcing piston down

Work done by piston converted to forward motion

Cruising

Air resistance converts kinetic energy to heat and sound

Kinetic energy converted to electrical energy stored in battery

Energy lost as heat and sound

Energy lost as heat and sound

Friction of wheels on road converts kinetic energy to heat

Electrical energy is transformed to light energy to illuminate office buildings at night. To power heating systems in cold weather, stored chemical energy (oil or gas) is burned, producing heat energy. The elevator motors are powered by electrical energy, which is turned into kinetic energy (energy of motion) as the elevators go up and down.

Deceleration

Brakes convert kinetic energy to heat and sound

Brake mechanism

Electrical energy used to work lights, wipers, radio and horn

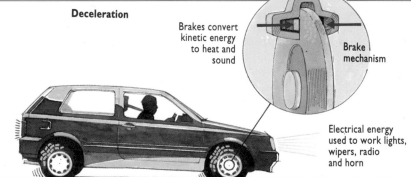

Earth's energy store

Much of the energy we use on Earth comes from sunshine stored away millions of years ago.

In our daily lives, energy is used in innumerable ways. For instance, we call on it every time a light is switched on or a car is driven from one place to another. But where does this energy come from? The electrical energy that powers the light probably came from a power station burning coal, oil, or gas, or from a nuclear power station, and the car's fuel came from oil.

All these sources of energy – coal, oil, gas, and the radioactive fuel in a nuclear power station – are stored forms of energy. Energy in radioactive elements, such as uranium and thorium, has been held in them since the heavy elements of the Earth were born in the explosion of a supernova deep in space, long before the solar system formed. The fossil fuels coal, oil, and gas gained their energy much later in geological history when plants or creatures used energy from the Sun to capture carbon from the air.

The energy sources on which we now depend have advantages and disadvantages. Fossil fuels are cheap to extract and convenient to use. But burning oil and coal can give unwelcome emissions and, along with burning gas, raises the carbon dioxide level in the atmosphere. Also, the stores of these fuels cannot be renewed, and so may run out one day. So far, new reserves have been found faster than these fuels have been used up, but there are fears that the end of oil and gas may not be far away – perhaps only a few decades. However, the reserves of coal known about today will last for several centuries.

Nuclear energy does not raise carbon dioxide levels and, under normal operating conditions in power stations, does not pollute the air in other ways.

*Coal is rich in the stored energy of sunlight. Coal (**above**) is the fossilized remains of plants that flourished tens or even hundreds of millions of years ago. When, about 300 million years ago, the fernlike trees of the Carboniferous period died and rotted, many were buried in mud. As time passed, they were chemically transformed, and in a few thousand years peat formed.*

By about 250 million years ago, overlying rock had compacted the peat to form lignite, or brown coal. This contains between 65 and 70 percent carbon. About 100 million years ago, much of this had turned into bituminous coal. With more time anthracite formed. This is made of up to 95 percent carbon and is valued today as the best type of coal for burning. To make 1 ton of coal takes between 25 and 75 tons of growing vegetation.

Coal is often extracted from mines over 3,000 feet (1 km) deep. But as ways are found of using coal of poorer quality, previously uneconomic shallow seams and near surface deposits are being exploited.

Dead vegetation forms peat

Peat slowly compressed into lignite

Further compression forms bituminous coal

Anthracite coal seam finally forms

300 million years ago

250 million years ago

100 million years ago

Today

But there have been leaks of radioactive materials. There are also fears of escapes of dangerous radioactive materials, such as happened at Chernobyl in the Ukraine in 1986. Scientists are researching into nonpolluting, renewable energy sources such as wind, wave, geothermal, and solar power.

Flames flare on an oil drilling platform as petroleum gases are burned off. Oil's stored energy accumulated long ago when sea creatures built up their tissues using energy from the Sun. When they died, they fell to the seabed and were covered with sediment. Under pressure and heat, oil was formed after millions of years.

Energy without limit

Abundant energy arrives at the surface of the Earth, but only a minute amount of it is captured and put to work.

Of the energy reaching the Earth's surface, 78 percent is sunlight, which powers plant growth and the weather, and helps create the winds and waves. Heat from inside the Earth gives a further 20 percent and the final 2 percent is gravitational energy, from the Moon and Sun, which raises the ocean tides. The energy from the sunlight hitting the Earth every day is so great that if it were collected, it could power the entire world's industry for several years. So far people have made little use of this and other sorts of free and virtually everlasting energy – the reason is that this energy is difficult to collect.

For instance, sunshine is thinly spread: if the energy from sunlight falling on about 1 square foot (900 cm^2) of the Earth's surface with the Sun directly overhead could be collected, it would be about enough to light only a 100W bulb. And wave power machines and power from hot rocks are only viable in places where there are reliable waves or where Earth's heat is available near the surface.

Energy from the Sun can be exploited both indirectly and directly. Indirect use relies on nature to collect it over a period of time during which the "biomass" sources of energy, such as firewood are created. Although sunshine can be used directly in solar homes for heating and hot water, solar cells that convert sunshine into electricity are expensive to make and not very efficient.

An energy source to rival power from conventional fuels is fusion power. But despite intense research, fusion power is still many years away from being at all practicable.

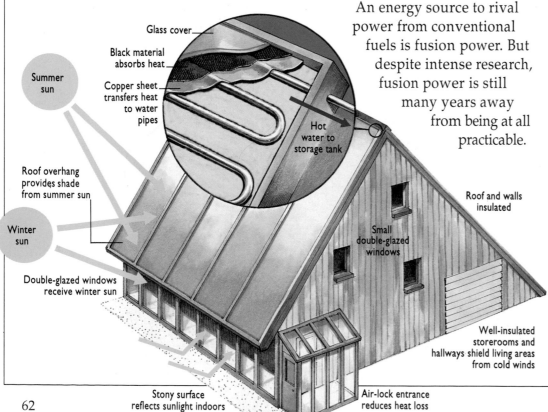

Glass cover

Black material absorbs heat

Copper sheet transfers heat to water pipes

Hot water to storage tank

Summer sun

Winter sun

Roof overhang provides shade from summer sun

Double-glazed windows receive winter sun

Roof and walls insulated

Small double-glazed windows

Well-insulated storerooms and hallways shield living areas from cold winds

Stony surface reflects sunlight indoors

Air-lock entrance reduces heat loss

The solar-heated home (above) captures energy from sunshine. Although it may be uneconomic to use solar energy in industry, the home is a practical place to exploit free energy from our local star.

Energy from sunlight is captured to heat the interior of the house and to heat water (left). The roof is angled to catch the maximum sun during the day, and heat is passed to water in pipes beneath the roof's outer layers.

In summer, when the Sun is high in the sky, the large roof overhang stops the interior from getting too hot. In winter, sunlight shining through the windows warms the interior. Parts of the house facing away from the Sun have small windows and are insulated to keep heat inside.

A forest of wind turbines harvests the energy of the wind at a wind farm in the Mojave Desert in California. An advanced turbine has huge turbine blades 180 feet (55 m) across and can develop up to one megawatt of electrical power.

For all intents and purposes, as long as the Sun shines, the wind will blow. This is because the wind redistributes the heat of the Sun from the hot equatorial regions to the cooler areas closer to the poles. So wind power will be available for several billion years to come.

Harnessing energy

*Take a mouthful of food, and you set off a series of energy changes
that gives your body the power to keep you alive.*

Energy cannot be created or destroyed. So to make use of it, energy has to be captured in its
original form or harnessed as it transforms itself from one form to another. For instance, the
stored energy of a fuel, such as coal, oil, or gas, does not heat a home until it is released when
the fuel is burned to make heat energy.

To harness energy or make use of a transformation, the energy has to be in a handy and
accessible form. One of the most useful forms of energy is electrical energy. To make it,
fuel is first burned, liberating stored energy to give heat. The heat is used to boil
water to make steam, which drives a generating turbine. The electricity made can
then be sent over huge distances, usually along cables suspended on pylons. At
its destination it can at last do useful work in the home or factory – heating,
lighting, driving motors, or powering television sets and home computers.

To get the energy we need to move about, grow, keep warm and
think, we eat food. The food is digested and the energy contained
within it released during countless complex chemical reactions
that take place throughout the body.

A waterwheel not only captures energy in its original

*As the wind powers the windsurfer across the water, its kinetic
energy (energy of motion) is converted to kinetic energy in the board,
sail, and surfer, and in the water pushed aside by the board. This is
the simplest kind of energy use – energy is transferred from one
place to another without a change of form.*

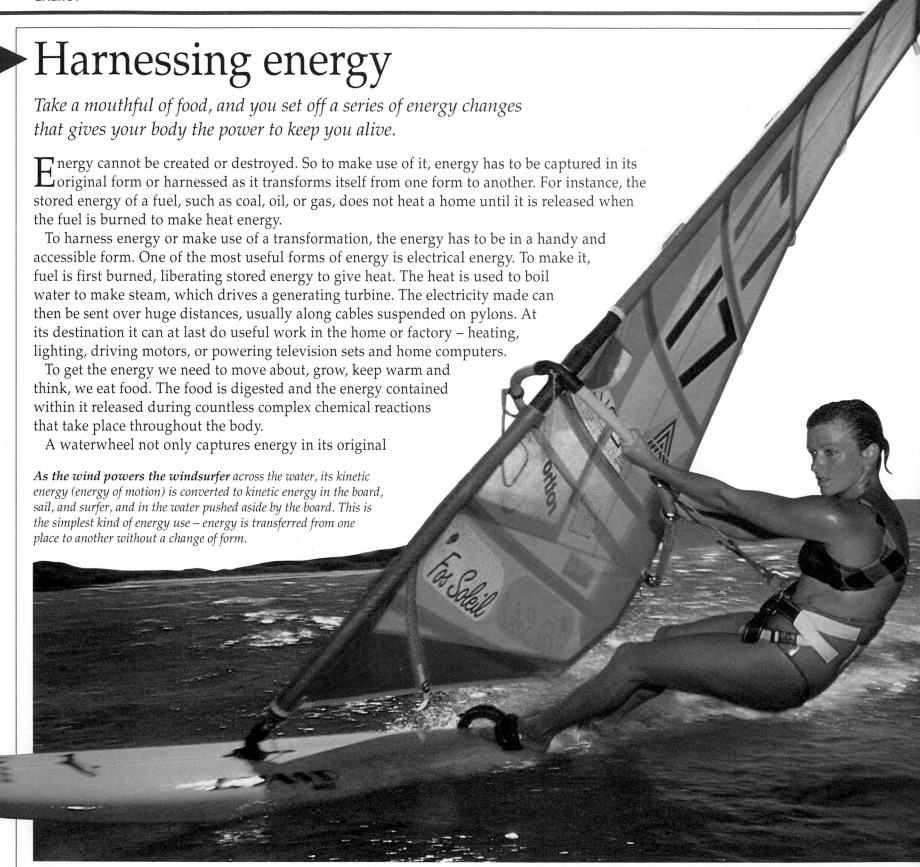

The Earth's internal heat is a pollution-free supply of energy that is, for all practical purposes, limitless. In Iceland heat from underground hot springs is used to drive power stations to give electrical energy. The hot water from the springs is also used to heat homes.

form, but also makes use of a transformation. A waterwheel taps both the kinetic (motion) and potential energy of flowing water. As the swiftly flowing water from a stream or channel hits the top of the wheel, the kinetic energy of the water is captured. The water collects in the wheel's paddles, and the potential energy of the weight of the water pulls down the wheel and is converted from potential to kinetic energy. The end result is rotary kinetic energy, which can be used to do useful work such as grinding corn. Its modern equivalent, the water turbine, also turns the kinetic energy of water into rotary motion, which in turn is converted into electrical energy on a vast scale in hydroelectric power plants.

See also

► **ENERGY**
  Making things move
  48/49

► Energy's endless flow
  58/59

► Earth's energy store
  60/61

► Energy without limit
  62/63

► Generating electricity
  68/69

SPACE
► Land, sea, and air
  14/15

ATOMS AND MATTER
► Heated reactions
  118/119

ENERGY FROM FOOD

Antelope (**right**) survive by running fast from predators. But energy for fast running cannot be obtained directly from their vegetable diet so the food is converted from one chemical form to another. It is stored in the antelope's body as the sugar glucose, or as glycogen, which can rapidly be turned into glucose. When energy is needed for a quick getaway, the sugar is rapidly "burned," breaking down as it combines with dissolved oxygen in the blood. The energy from food also generates heat and maintains the many processes in the creature's body (**below**).

Food energy

Released by respiration and digestion

Used moving

Used for growth and repair

Lost in waste products

Lost as heat

Sparks and attractions

The energy to make a light bulb shine comes from movements of tiny particles. These same particles make a compass needle point north.

Both electricity and magnetism depend on the transfer or movement of electrons – minute negatively charged particles that make up the outer regions of atoms. When electrons move from atom to atom – down a wire, for instance – the movement produces a flow of electric current. But what makes electrons flow? The answer depends on the fact that like charges repel and opposite charges attract each other. When an object has an excess of electrons, for one reason or another, it is negatively charged, and when an object lacks electrons, it is positively charged. Thus electrons, which have a negative charge, are drawn toward a positively charged object to correct the balance and are repelled by a negatively charged object. The space between two objects that have different charges is called a field, and electrons always move, if they can, toward the positive side of the electric field.

If hair and a plastic comb are rubbed together, the hair tends to stick to the comb because an electric field is created. A relatively small number of negatively charged electrons has been shifted from the hair to the comb, so the hair has an overall positive charge. Since the comb has an excess of electrons, it is negatively charged, and the hair and comb are attracted to each other.

Electron movement also causes natural magnetism in certain minerals, such as magnetite. When atoms are in the regular lattice structure of a mineral, they line up and point in the same direction. The total effect of all the electrons swirling around the atomic nuclei is like that of a current flowing around the mineral. Since magnetic and electric fields are directly linked, as soon as an electric field changes, or electrons move and a current flows, a magnetic field at right angles to the electric field, or current, is created. So, because the electrons create a current while they are orbiting the atoms in the mineral, a magnetic field is created. The link also operates in reverse, so that when a magnetic field varies in strength or direction, it produces an electric field.

All magnets have two regions, called poles, in which the magnetic force seems to be concentrated. The Earth itself is like a giant magnet, with poles of its own. One pole of any magnet is attracted to the Earth's north magnetic pole, the other to the south. Like poles repel; opposite poles attract.

Electrical energy, like any other form of energy, can be transformed into other sorts of energy. In a light bulb, for instance, light comes from a wire filament. The energy of the electric current flowing through the wire is carried by electrons moving at high speed. The electrons moving through the wire jostle the atoms of the metal, making them vibrate faster and heating them up. Some of the energy that the atoms in the filament pick up is given off as electromagnetic radiation – both as visible light and invisible infrared (heat) radiation.

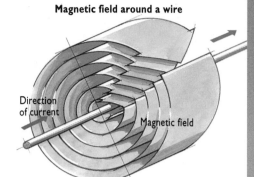

Magnetic field around a wire

Direction of current

Magnetic field

Magnetic field around a coil

Direction of current

Coiled wire

Magnetic field

*When an electric current flows, a circular magnetic field is always generated at right angles to the direction of the current (**top**).*

*If wire is coiled to make a solenoid (**above**), the field created is the sum of the fields of each turn of wire and is like the field of a bar magnet.*

*Electromagnetism is put to work in a special crane that lifts iron and steel scrap (**right**). The magnetic field of the crane's head is created by electric currents flowing in wires wrapped around an iron core. The iron core is magnetized when the current is switched on.*

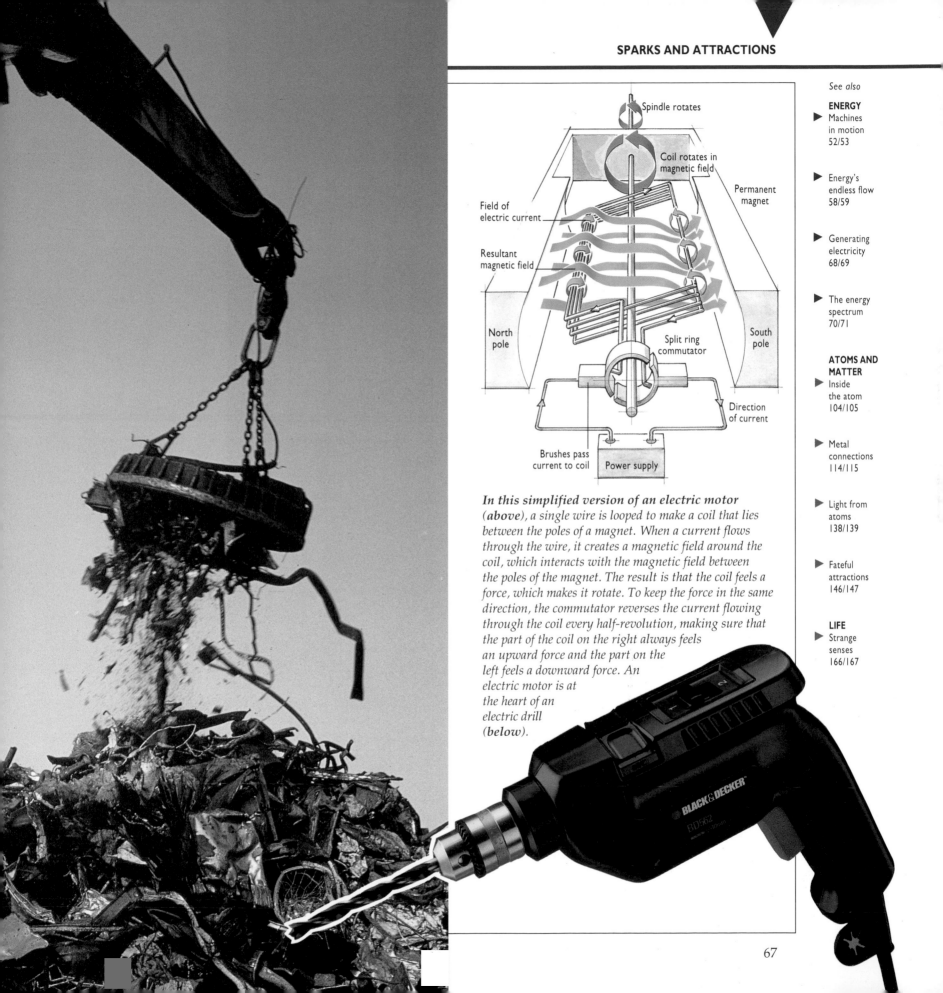

Spindle rotates

Coil rotates in magnetic field

Permanent magnet

Field of electric current

Resultant magnetic field

North pole

South pole

Split ring commutator

Direction of current

Brushes pass current to coil

Power supply

In this simplified version of an electric motor
*(**above**), a single wire is looped to make a coil that lies
between the poles of a magnet. When a current flows
through the wire, it creates a magnetic field around the
coil, which interacts with the magnetic field between
the poles of the magnet. The result is that the coil feels a
force, which makes it rotate. To keep the force in the same
direction, the commutator reverses the current flowing
through the coil every half-revolution, making sure that
the part of the coil on the right always feels
an upward force and the part on the
left feels a downward force. An
electric motor is at
the heart of an
electric drill
(**below**).*

BLACK & DECKER

BD562

Generating electricity

It is easy to take electricity for granted, but if you have ever wondered where it comes from, there is a simple explanation.

In homes and factories around the world, electricity is one of the most useful forms of energy. In fact, electrical energy is the workhorse of industry because it can easily be distributed and turned into other, useful forms of energy. Electricity is generated when coils of wire are rotated in a magnetic field. Turning the coils in the field creates an electromotive force, or EMF (measured in volts), which forces electrons along the wire to create a current. In power stations the force to rotate the coils is usually provided by turbines turned by high-pressure steam. The steam is made using heat from the burning of coal, oil, or gas, or from the fission (splitting) of radioactive materials such as uranium. In hydroelectric plants, the force comes from falling water. On a small scale, a bicycle dynamo, turned by the wheel, powers the cycle's lights.

The current from a power station generator is converted by transformers to a high voltage before being sent over long distances. By raising the voltage, the current goes down; the smaller the current, the smaller the losses caused by heating of the wires carrying the current. Electricity is sent at around 400,000 volts (400 kilovolts or 400 kV). After transmission, and close to where the electricity is going to be used, transformers lower the voltage again.

When it is not possible to plug into electricity, batteries can give small amounts of electrical energy. In all batteries, two materials – typically metals such as zinc and manganese – react together in the presence of a third substance called an electrolyte. When the reacting materials are connected by an external wire, a chemical reaction forces electrons along the wire and a current flows which can power electrical devices.

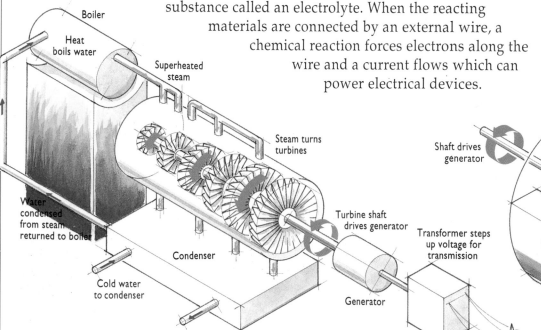

Boiler

Heat boils water

Superheated steam

Steam turns turbines

Water condensed from steam returned to boiler

Cold water to condenser

Condenser

Turbine shaft drives generator

Generator

Transformer steps up voltage for transmission

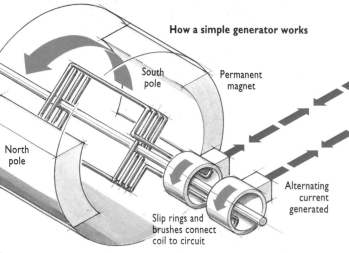

How a simple generator works

Shaft drives generator

South pole

Permanent magnet

North pole

Alternating current generated

Slip rings and brushes connect coil to circuit

High voltage cables

PORTABLE POWER

In a long-life alkaline battery, zinc powder in the central chamber dissolves in potassium hydroxide. The outer cylinder consists of manganese dioxide mixed with graphite. When the zinc dissolves, it gives up electrons which travel around the circuit. The electrons return to the battery via the casing and are taken in by the manganese dioxide.

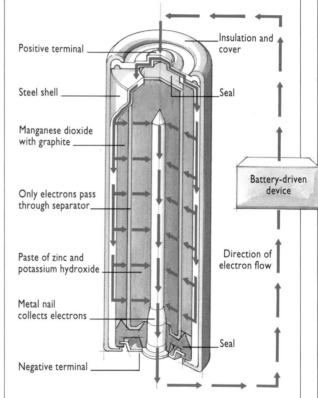

Positive terminal

Insulation and cover

Steel shell

Seal

Manganese dioxide with graphite

Only electrons pass through separator

Battery-driven device

Paste of zinc and potassium hydroxide

Direction of electron flow

Metal nail collects electrons

Seal

Negative terminal

Boiling water makes steam, which drives turbines around. *After turning the turbines, the steam condenses back to water. The turbine turns coils of wire in the strong magnetic field of a generator. Because the coils rotate in a magnetic field the current produced in them alternates – that is, it repeatedly changes its direction from positive, to negative, to positive, and so on.*

More than two-thirds of the energy stored in a fossil fuel is lost – mostly as heat energy – when electricity is generated. Most is lost in the cooling towers where the steam is condensed back into water after it has turned the turbines. Much is also converted to heat energy by friction in the turbines and generator. The rest is lost during transmission.

Energy from fuel 100%

Energy lost up chimney

Energy lost as friction in generator

Energy lost as heat and sound in transformers

Energy lost in turbines and condenser 65%

Energy lost from cables

Energy available for industrial and household electricity 30%

The energy spectrum

Listening to the radio, looking around, and getting a tan all use a fundamental form of energy – electromagnetism.

Electromagnetic waves spread through space like ripples across a pond when a stone is thrown in. Through empty space, all electromagnetic waves travel at about 186,000 miles per second (300,000 km/s). This is fast enough to circle Earth in one-seventh of a second or to travel the 93 million miles (150 million km) from the Sun in about eight minutes. Through matter, such as air or water, electromagnetic radiation travels more slowly – the denser the matter, the slower the speed.

Wavelength shortens toward violet end of spectrum

Visible light ranges from blue, with waves about 1.5 x 10⁻⁵ inch (3.8 x 10⁻⁷ m) long, to red, with waves up to 3 x 10⁻⁵ inch (7.5 x 10⁻⁷ m) long. Earth's atmosphere lets visible light through, while many other wavelengths of sunlight are blocked. Our eyes work using light because these wavelengths are available on the Earth's surface .

Wavelength in meters											Visible light		
Gamma rays used to treat cancer						X-rays		X-ray image		Ultraviolet rays tan skin		Infrared heat image of body	
Gamma rays										Ultraviolet		Infrared	
10⁻¹⁶	10⁻¹⁵	10⁻¹⁴	10⁻¹³	10⁻¹²	10⁻¹¹	10⁻¹⁰	10⁻⁹	10⁻⁸	10⁻⁷	10⁻⁶	10⁻⁵		

It is, in fact, the link between electricity and magnetism that is responsible for light and all the other radiations of the electromagnetic spectrum, including radio waves, X-rays, and microwaves. Electromagnetic radiation is always produced when an electron in an atom jumps from one orbit to another closer to the nucleus. The link exists because electromagnetic radiation is made up of both electrical and magnetic energy in about equal amounts, and electromagnetic radiation spreads across the universe as interacting waves of electric and magnetic fields. The bigger the jump made by the electron as it changes orbits, the more energy is carried by the resulting electromagnetic radiation, and the shorter its wavelength.

Capture and compare radiation from anywhere along the electromagnetic spectrum, and the only differences observable are in wavelength, frequency, and energy. These differences change the effects the waves have. For instance, short-wavelength, high-energy X-rays can penetrate our bodies while longer-wavelength, lower-energy visible light cannot.

Each wavelength of radiation can deliver a fixed amount of energy called a quantum. Electromagnetic radiation is thus also described in terms of "packets" of energy or particles. The particles of radiation are called photons; the shorter the wavelength, the higher the photon's energy.

Electromagnetic radiation exists in a continuous spectrum (above) of wavelengths from the very short to the very long. The shortest are gamma rays, made in radioactive decay, with wavelengths of less than 10⁻⁹ inch (3 x 10⁻¹¹ m). Their frequencies are higher than about 10¹⁹ hertz (cycles per second). The longest are ELF (extremely low-frequency) radio waves, with wavelengths thousands of miles long and frequencies of less than 300 hertz.

The rainbow (far right) is the sliver of the electromagnetic spectrum occupied by visible light. Raindrops are tiny prisms which can act together to refract sunlight and split it into the familiar colors of the rainbow. The longer the wavelength of the color, the less it is refracted, or bent, as it passes through denser matter – in this case water. Because the colors are bent by differing amounts, they are dispersed, or spread out, to form the colors of the rainbow.

See also

ENERGY
► Sparks and attractions
66/67

► Visible energy
72/73

► Beyond violet
76/77

► Over the rainbow
78/79

► A matter of degree
80/81

WAVES OF ENERGY

Electromagnetic radiation is made up of rapidly changing electric and magnetic fields. At each point along the wave, there is an electric and magnetic field. Each is at right angles to the direction of movement of the wave. Along the wave, each field goes up and down in strength. The distance between one high point and the next in either the electric or magnetic field is known as the wavelength of the radiation. The number of high points that pass by every second gives the frequency of the wave, which is measured in hertz (abbreviated to Hz), or cycles per second. In the wave shown, the electric field is vertical and the magnetic field horizontal. In general, the fields vary in direction, but always stay at right angles to the direction of the radiation.

Wavelength lengthens toward red end of spectrum

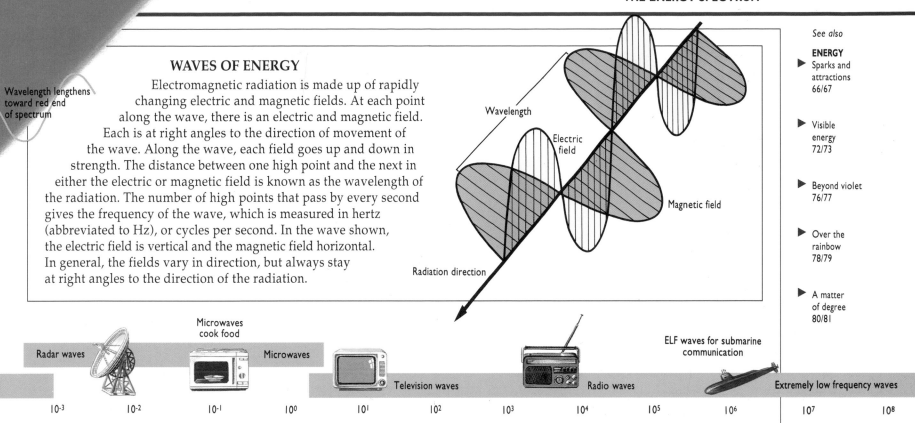

Wavelength

Electric field

Magnetic field

Radiation direction

Radar waves

Microwaves cook food

Microwaves

Television waves

Radio waves

ELF waves for submarine communication

Extremely low frequency waves

10^{-3} 10^{-2} 10^{-1} 10^{0} 10^{1} 10^{2} 10^{3} 10^{4} 10^{5} 10^{6} 10^{7} 10^{8}

SPACE
► Our local star
22/23

ATOMS AND MATTER
► Inside the atom
104/105

► Light from atoms
138/139

► Fateful attractions
146/147

LIFE
► Using light
160/161

BRAINS AND COMPUTERS
► Window on the world
194/195

Visible energy

Light can be bent, bounced, and split. It can even make images we can see that are not really there at all.

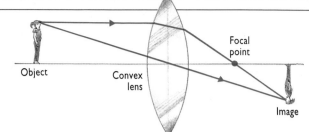

Of the vast amounts of radiation emitted by the Sun, only a small amount actually gets down to the Earth's surface. Part of this small "band" of radiation is light.

The fact that light generally travels in straight lines enables us to tell where objects are when our eyes detect the light reflected from them. Our sense of sight also depends on the fact that light can form images. Images are made when light is bent onto a new path, either by being reflected from a smooth surface, such as a mirror, or by being bent or refracted when it passes from one transparent medium to another, such as from air to glass.

Refraction happens because light slows down when it enters a more dense medium, and speeds up when it leaves. A lens refracts light in such a way that it bends light rays to form images. The human eye uses a lens to help focus images on the retina, a sensitive layer at the back of the eye. Light rays are focused mainly by the curved transparent front of the eye, the cornea. The fine adjustment needed to make a sharp image comes from changes in shape of the lens itself, which fine-tunes the refraction.

At the shiny reflecting surface of a mirror, light bounces off at an angle equal to the angle at which it hits the surface. Light seems to spread from the position, either behind or in front of the mirror, where the source of the light seems to be.

Light waves of different wavelengths all behave in the same way when they are reflected. But this is not so when they are refracted. Red light is bent least, violet light most. The different colors can be separated when

*In a convex lens light rays from an object converge or bend inwards, and form an image on the opposite side of the lens. It is a real image because it can be focused onto a screen. If the object is farther away than the focal point of the lens (**above**), the image is upside down. It needs magnifying by other lenses to give us binoculars or telescopes. But if the object is nearer than the focal point, the image is then upright and larger, as in a magnifying glass.*

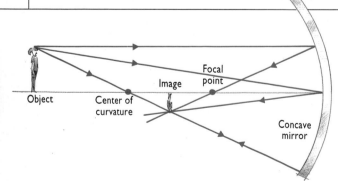

*An upside-down image appears in a concave mirror (**above**), such as the inside of the bowl of a spoon (**right**). Light rays from someone looking into the spoon are reflected, then cross over. The result is that, viewed from a little distance, the rays appear to come from a place in front of the spoon.*

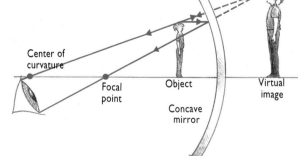

*Concave mirrors can magnify. If an object is close to a concave mirror (**above**), then the light rays do not cross over and are not actually brought to a focus on the reflecting side of the mirror. Instead, they form a virtual image, which is larger than the object, on the other side of the mirror.*

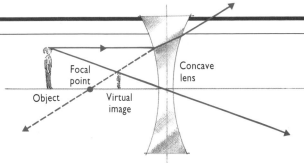

A **concave lens** (**above**) bends light away from its center line. It always creates an image that appears to be on the same side of the lens as the object and is smaller than the object. Because the light rays do not really come from the position at which the image seems to be, it is known as a virtual image. Near-sighted people use glasses with concave lenses to correct their vision and focus sharp images on the retina.

Seen through the side of a glass of water (**right**), a straw appears bent, shifted, and magnified, because light rays are themselves bent, or refracted, as they pass from water to glass, and from glass to air.

Points farther from the center of the glass seem shifted more than those closer to the center, causing the straw to look "fatter" than it really is.

Light rays of different colors are also bent differently: red least, violet most. Images formed by refraction often have colored fringes as a result. The effect is clearest if a narrow ray of white light is passed through a triangular prism.

light is refracted in glass, raindrops, and the lenses of optical instruments, to yield rainbow colors. Many machines make use of the fact that light can be bent and reflected. Just one of these is the camera, which is designed to form a sharp, undistorted image on a piece of light-sensitive film. On the film the energy from the light triggers a chemical reaction. When the film is developed, a permanent record of the image is imprinted and can be printed on paper.

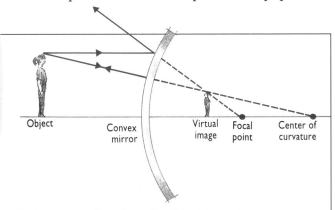

An image smaller than the object forms in a convex mirror, such as the underside of the bowl of a spoon (**left**). After reflection from the spoon, light rays from someone looking into the spoon appear to spread out from a place behind the spoon. A smaller image of the person appears at the place. It is the right way up because the reflected light rays do not cross over.

See also

ENERGY
▶ The energy spectrum 70/71

▶ The light of science 74/75

▶ Beyond violet 76/77

▶ Over the rainbow 78/79

SPACE
▶ The giver of life 24/25

ATOMS AND MATTER
▶ The atomic paintbox 136/137

▶ Light from atoms 138/139

LIFE
▶ Using light 160/161

BRAINS AND COMPUTERS
▶ Window on the world 194/195

The light of science

From playing music to fighting wars, lasers abound. What makes them unique and how do they work?

Lasers are used in compact disk players, supermarket checkout bar-code readers, delicate surgical procedures, and countless other situations where a source of light needs to be compact yet intense. Lasers are even used in weapons of war, guiding bombs to their targets with pinpoint accuracy.

The name "laser" stands for "Light Amplification by Stimulated Emission of Radiation." The laser effect takes place when a photon – a "particle" of light – hits an atom that has an excess of energy. The photon stimulates the emission of an identical photon, which then hits another atom and in turn generates another photon. Eventually a cascade of photons is produced, and the light is all at the same wavelength.

Many different types of laser have been built. Depending on what the laser is made of, it can send out light at wavelengths ranging from ultraviolet to infrared, with beams created continuously or in flashes. Some lasers deliver energy at about the same rate as a pocket calculator uses up its batteries; others deliver a thousand times the power of an electricity generating station – but in such a brief flash that it does not devour an enormous amount of energy.

Laser light is unique in that it is coherent – its waves are all in step with each other so its energy is concentrated into a very narrow beam. In normal light the waves are out of step so their energy is spread over a large area.

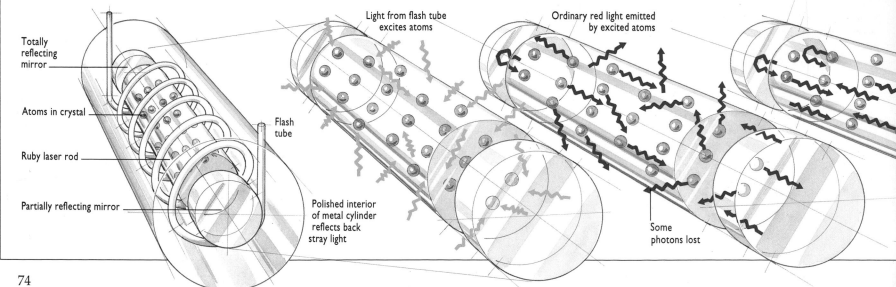

Totally reflecting mirror

Atoms in crystal

Ruby laser rod

Partially reflecting mirror

Flash tube

Polished interior of metal cylinder reflects back stray light

Light from flash tube excites atoms

Ordinary red light emitted by excited atoms

Some photons lost

Laser displays at pop concerts
(*left*) add a high-tech visual
dimension to the atmosphere.

At the heart of this laser is a
synthetic ruby crystal. The "active"
material is chromium, present in
the ruby as an impurity, giving it
its deep red color. When flooded
with light from a conventional flash
tube, chromium atoms are boosted
to a higher energy level. They shed
this energy as photons of ordinary
red light. The few photons that
happen to travel along the axis of
the crystal are reflected from the end
mirrors. As they traverse the crystal,
they stimulate the emission of further
photons, traveling in the same
direction and with the same phase.
These are also reflected. A pulse of
light builds up and then "leaks"
away through one end-mirror,
which is slightly transparent.

Laser light waves are of a single
definite wavelength and are all in
phase and parallel to each other
Ordinary light, by contrast, has a
spread of wavelengths and consists
of "wave-packets" that are "out of
step" with each other. Laser light
consists of long trains of waves that
are both intensely powerful and
form a parallel beam
that does not
spread out.

See also

▶ **ENERGY**
The energy
spectrum
70/71

▶ Visible
energy
72/73

▶ **ATOMS AND
MATTER**
Solid
structures
96/97

▶ Inside
the atom
104/105

▶ Light from
atoms
138/139

▶ Exotic
particles
144/145

▶ **BRAINS AND
COMPUTERS**
Beyond
the chip
232/233

Photons within rod
stimulate other
already excited atoms

Cascade of photons builds up

Ordinary red light
contains several
different wavelengths
which are out of step

Intense beam
of pure red light
emitted by laser
through partially
reflecting mirror

A few photons
leak out

Most photons
reflected back

Beyond violet

What makes the radiation beyond the violet end of the rainbow both useful and dangerous, and how do we "see" with invisible rays?

On either side of visible light – the rainbow colors of the spectrum – there is electromagnetic radiation that we cannot see. On the far side of violet, the wavelengths of the radiation get progressively shorter. As they do, the radiation becomes more energetic, can penetrate deeper, and is more harmful. After visible violet comes ultraviolet (UV), then X-rays, and finally gamma rays.

UV is radiated by extremely hot, powerful sources such as the Sun. Most UV is filtered out by the ozone layer in the atmosphere and so does not reach the Earth's surface. This is just as well because, although most UV does not have much penetrating power, too much can damage growing plants and burn skin. UV can be produced "artificially" by passing an electric current through a low-pressure gas; this is what happens in fluorescent lights, in which a coating on the inside of the tube then absorbs UV and in its place gives off the energy as visible light.

Powerful X-rays are made when high-energy electrons slam into atoms. Like UV, X-rays are produced in the Sun. On Earth they can be made by beaming electrons at a metal target in an X-ray machine. The X-rays produced are used in medicine to "see" inside the body.

Gamma rays are the most powerful and penetrating in the electromagnetic spectrum. Most are released when atoms break down during nuclear reactions. Although exposure to gamma rays can damage cells in the body, controlled gamma radiation is used to treat some cancers.

Sunbathers are bombarded with ultraviolet (UV), a potentially harmful form of electromagnetic radiation. Although only a fraction of the UV the Sun emits reaches the Earth's surface, it is enough to trigger one of the body's defense mechanisms. To protect itself, the skin secretes melanin, a dark substance which causes tanning. Too much sun overloads the system, leading to severe sunburn and even cancer, a particular risk for fair-skinned people.

Flowers lure bees using ultraviolet patterns. Bees can see UV radiation that human eyes cannot detect. Flowers have evolved to make use of the bees' ability and have markings on their petals that are invisible to us. The bees use these markings as a runway to guide them in to the flower's nourishing nectar. The flower gains because it is pollinated by the bees.

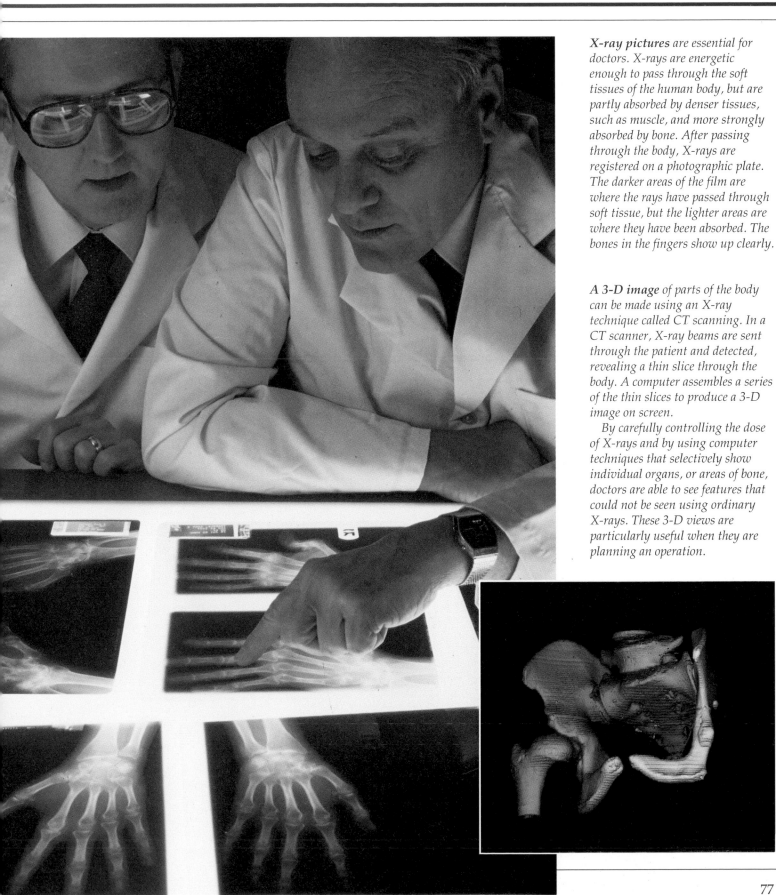

X-ray pictures are essential for doctors. X-rays are energetic enough to pass through the soft tissues of the human body, but are partly absorbed by denser tissues, such as muscle, and more strongly absorbed by bone. After passing through the body, X-rays are registered on a photographic plate. The darker areas of the film are where the rays have passed through soft tissue, but the lighter areas are where they have been absorbed. The bones in the fingers show up clearly.

A 3-D image of parts of the body can be made using an X-ray technique called CT scanning. In a CT scanner, X-ray beams are sent through the patient and detected, revealing a thin slice through the body. A computer assembles a series of the thin slices to produce a 3-D image on screen.

By carefully controlling the dose of X-rays and by using computer techniques that selectively show individual organs, or areas of bone, doctors are able to see features that could not be seen using ordinary X-rays. These 3-D views are particularly useful when they are planning an operation.

Over the rainbow

Unseen electromagnetic waves heat our food and carry messages around the world and into space. The length of the waves determines how they can be used.

Beyond the color red in the electromagnetic spectrum lie radiations that have longer wavelengths and lower energies than those of visible light. Those immediately adjacent to the visible spectrum are called infrared, or "below-red," radiations. Infrared radiation is often known as heat radiation because, at everyday temperatures, the bulk of the invisible radiation given off by objects around us as heat is in this range. Infrared waves can jostle atoms and molecules and set them moving – that is, they heat up.

Humans and many other animals can feel heat radiation when it hits the skin, but some creatures actually "see" with infrared. The pit viper is named after the pits on each side of its nostrils that are sensitive to infrared radiation given out by prey animals such as mice. By turning and tilting its head, the snake locates the mouse and strikes in total darkness.

Infrared ends at a wavelength of about ⅟₂₅ inch (1 mm) which is 1,500 times the wavelength of red light. Beyond it lies

the microwave band, with wavelengths up to 12 inches (30 cm). Microwaves do not generally have the strong heating effects of infrared wavelengths, but some happen to be of the right frequency to set molecules of water in motion – an effect used in microwave ovens. Microwaves are the highest frequencies of radio waves. Tight beams of them, focused by dish-shaped aerials mounted on towers or hills, transmit telephone communications. Radar makes use of microwaves and short-wavelength radio wavelengths from ⅖ to 39 inch (1 to 100 cm).

Microwaves, and the radio waves that lie beyond them, set electrons in motion and generate small electric currents when they hit conducting materials such as copper wire. This is what makes radio communication possible. The small electric currents which are received are amplified and used, for instance, to power loudspeakers and control TV images.

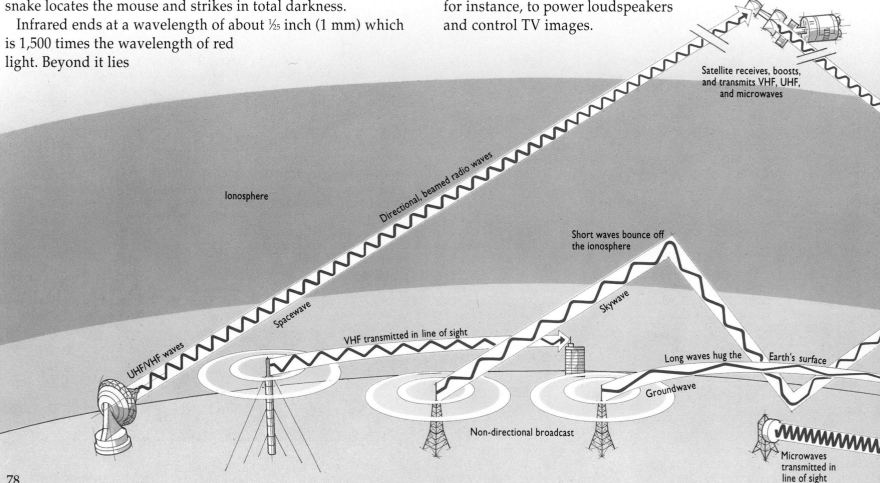

Satellite receives, boosts, and transmits VHF, UHF, and microwaves

Ionosphere

Directional, beamed radio waves

Short waves bounce off the ionosphere

Spacewave

Skywave

UHF/VHF waves

VHF transmitted in line of sight

Long waves hug the   Earth's surface

Groundwave

Non-directional broadcast

Microwaves transmitted in line of sight

Satellites can observe the heat radiation given out by objects on Earth. The picture (**left**) of San Francisco was taken from Landsat with infrared-sensitive film. At these wavelengths, the familiar features of the city show up clearly. The Golden Gate Bridge, across the entrance to the bay, is at upper left.

MICROWAVE HEATING

Each water molecule is made up of an oxygen atom linked to two hydrogen atoms, which are positioned at an angle, forming "ears." Water molecules are polar molecules, so in an electric field they line up: if the field is vertically upward, the "ears" tend to be upward, too. In a microwave, food is bathed in microwaves, typically with a frequency of 2.45 gigahertz (2.45 billion oscillations per second). As the electric field component of the radiation changes polarity, water molecules (**below**) line up in one direction and then swivel and reverse direction in time with the field. The friction between the rapidly swiveling water molecules and the food molecules speeds up vibrations of all molecules so the food heats up.

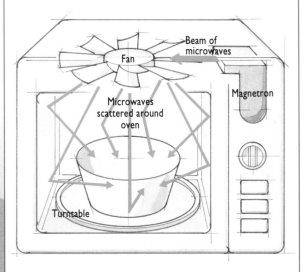

Beam of microwaves

Fan

Magnetron

Microwaves scattered around oven

Turntable

Different radio waves of the electromagnetic spectrum have different properties. Relatively long wavelength waves, such as in the long, medium, and short wave bands, with wavelengths ranging from roughly 6 miles to 33 feet (10 km to 10 m) can be used to broadcast radio signals over long distances. Long waves do this by bending around the Earth to produce groundwaves. Short waves do this by being reflected off a layer in the atmosphere called the ionosphere, creating skywaves.

Shorter wavelength radio waves such as VHF (very high-frequency), UHF (ultrahigh-frequency), and microwaves, neither curve around the Earth nor bounce off the ionosphere so they can be used only to broadcast locally in "line of sight" carrying TV, FM radio, and telephone communications. But since they penetrate the ionosphere, they can be relayed around the world by communication satellites orbiting high up in space.

Positive end of
water molecule

Negative end of
water molecule

Water molecule jiggles about
as microwave passes

Changing field
of microwave

A matter of degree

When you feel how hot your bath water is or take your temperature, what exactly are you measuring?

Heat is basically the movement of molecules – the tiny particles that make up every substance. When you put your hand over a heater, the warmth you feel is simply an assault by billions of fast-moving molecules. The faster they are moving, the hotter the heater is.

We measure how hot something is in terms of temperature. But temperature is not the same as heat. Heat is energy, the combined energy of all the moving molecules; temperature is simply a measure of how fast, on average, they are moving.

Heat energy can be converted into other forms of energy and vice versa. A steam engine, for instance, turns heat energy into mechanical energy; rubbing your hands together turns mechanical energy into heat. But the relationship between heat and temperature is subtle.

Heating a substance usually makes it hotter – that is, it raises its temperature – but by how much varies with the substance as well as the amount of heat. For instance, it takes about twice as much heat to raise the temperature of a pan full of water as it does to raise the temperature of a pan half full of water by an equal amount. But with equal weights of iron and lead, it takes three times the heat to raise the temperature of the lead by one degree since more energy is needed to set the heavy molecules of lead vibrating by the same amount.

Heat does not always make things hotter. When a substance is melting or boiling, adding more heat does not raise the temperature; it simply makes more of the substance melt or boil. If you heat melting ice, for instance, the ice melts faster, but its temperature stays at 32°F (0°C) – because the molecules of ice do not move any faster. All the heat energy goes into breaking the bonds between molecules and turning the ice to water. Similarly, boiling water is always at 212°F (100°C) under normal conditions, no matter how much you heat it.

Sparks flying *from a sanding wheel are vivid evidence of the way mechanical energy – the movement of the wheel – can be turned into heat energy. The mechanical energy came in turn from the electrical energy driving the grinder's electric motor. The pieces of glowing metal that make up the sparks are at a high temperature, but do not actually have much heat because they are light and tiny.*

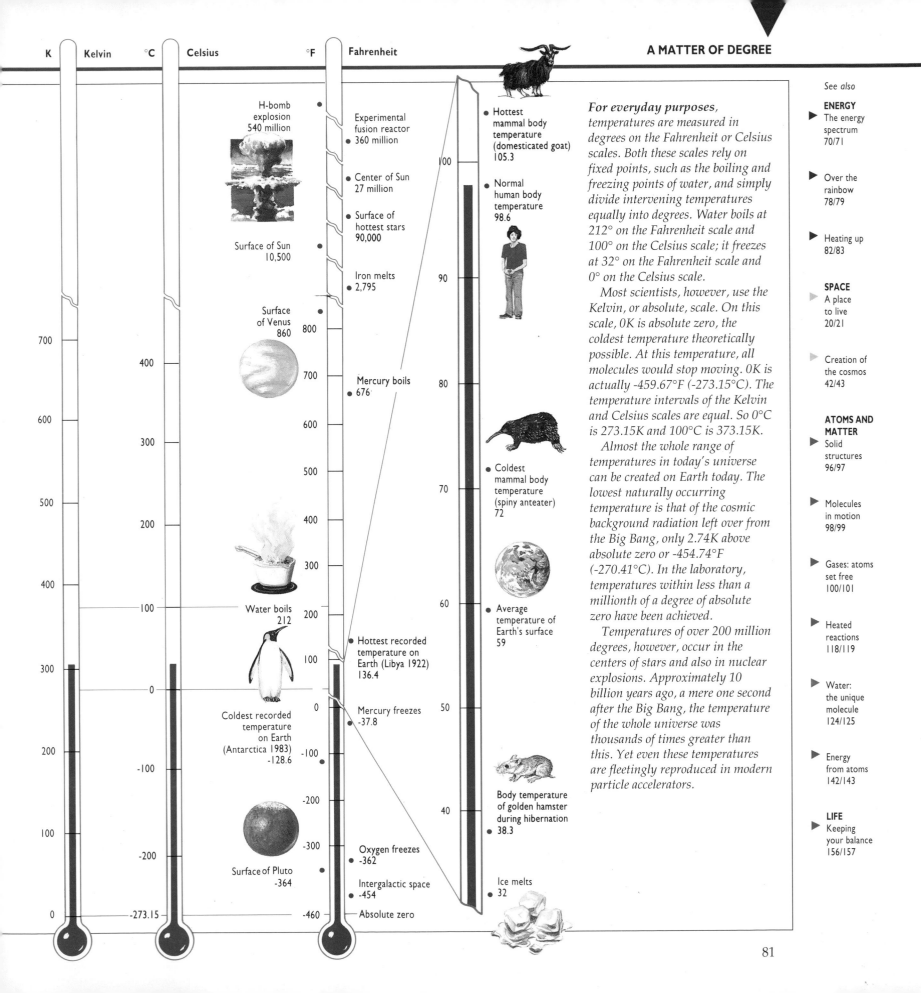

K	Kelvin	°C	Celsius	°F	Fahrenheit

H-bomb explosion 540 million

Experimental fusion reactor 360 million

Hottest mammal body temperature (domesticated goat) 105.3

Center of Sun 27 million

Normal human body temperature 98.6

Surface of hottest stars 90,000

Surface of Sun 10,500

Iron melts 2,795

Surface of Venus 860

Mercury boils 676

Coldest mammal body temperature (spiny anteater) 72

Average temperature of Earth's surface 59

Water boils 212

Hottest recorded temperature on Earth (Libya 1922) 136.4

Mercury freezes -37.8

Coldest recorded temperature on Earth (Antarctica 1983) -128.6

Body temperature of golden hamster during hibernation 38.3

Surface of Pluto -364

Oxygen freezes -362

Intergalactic space -454

Ice melts 32

Absolute zero

For everyday purposes, *temperatures are measured in degrees on the Fahrenheit or Celsius scales. Both these scales rely on fixed points, such as the boiling and freezing points of water, and simply divide intervening temperatures equally into degrees. Water boils at 212° on the Fahrenheit scale and 100° on the Celsius scale; it freezes at 32° on the Fahrenheit scale and 0° on the Celsius scale.*

Most scientists, however, use the Kelvin, or absolute, scale. On this scale, 0K is absolute zero, the coldest temperature theoretically possible. At this temperature, all molecules would stop moving. 0K is actually -459.67°F (-273.15°C). The temperature intervals of the Kelvin and Celsius scales are equal. So 0°C is 273.15K and 100°C is 373.15K.

Almost the whole range of temperatures in today's universe can be created on Earth today. The lowest naturally occurring temperature is that of the cosmic background radiation left over from the Big Bang, only 2.74K above absolute zero or -454.74°F (-270.41°C). In the laboratory, temperatures within less than a millionth of a degree of absolute zero have been achieved.

Temperatures of over 200 million degrees, however, occur in the centers of stars and also in nuclear explosions. Approximately 10 billion years ago, a mere one second after the Big Bang, the temperature of the whole universe was thousands of times greater than this. Yet even these temperatures are fleetingly reproduced in modern particle accelerators.

Heating up

When a kettle is boiled, how does the water gain its heat? And when the Sun warms us up, how does the heat travel through space?

Heat energy in vibrating atoms or molecules can be passed on and made use of in many ways. The two main ways in which heat can be transferred are conduction and radiation. Heat is also moved about by convection.

Wherever atoms or molecules touch, energy is conducted, whether in solids, liquids, or gases. Inside a coffeemaker, for instance, the atoms in the heating element start to vibrate vigorously as the machine heats up. In solids, where particles are tightly packed, the vibration of one particle immediately starts its neighbor moving, so heat spreads fast.

The metal atoms in the element touching water molecules bang into them, making them vibrate more, thus raising their temperature. A vigorously vibrating molecule needs more space, so the hot water becomes less dense and floats upward. To replace the rising hot-water molecules, cool water is drawn in toward the element, where it heats up, expands, and moves upward. As they rise, the hot molecules "give" energy to molecules they bang into. They cool, shrink, and become denser until they match the density of the water around them.

This heat transfer from place to place by moving particles is convection; it goes on only in liquids or gases.

Vibrating particles (atoms or molecules) always give off electromagnetic radiation. The heat we feel from the Sun has been

In a pan of boiling water heat is transferred in three ways. Heat is conducted from the electric ring to the pan sitting on it. This energy then moves, again by conduction, into the water molecules touching the pan. These vibrate, expand and rise, carrying heat by convection. When water boils, more heat is convected upward in the steam. Unlike conduction and convection, where heat is carried along physically by molecules, heat also travels as radiation. The electric ring gives off visible energy as it glows red, but both ring and pan radiate far more heat as invisible infrared radiation.

The white paint on houses on the Greek island of Thira keeps them cool in summer by reflecting the Sun's heat. Because white light is made up of all the colors of visible light, a white surface reflects back all wavelengths, thus absorbing far less energy than other colors. This is why people tend to wear pale colors in summer and darker, heat-absorbent, colors in winter.

See also

▶ **ENERGY**
Energy's
endless flow
58/59

▶ The energy
spectrum
70/71

▶ Over the
rainbow
78/79

▶ A matter
of degree
80/81

SPACE
▷ Land, sea,
and air
14/15

▷ Our local
star
22/23

▷ The giver
of life
24/25

**ATOMS AND
MATTER**
▷ Solid
structures
96/97

▷ Molecules
in motion
98/99

▷ Gases: atoms
set free
100/101

LIFE
▷ Keeping
your balance
156/157

radiated by the energetic vibrating particles that make up the Sun. Radiation can travel through the vacuum of space and, unlike conduction or convection, does not need the intervention of matter for its transfer.

When radiation hits an object, energy is passed on which excites the molecules and makes them vibrate faster; their temperature rises. Fast-moving, high-temperature molecules give off high-energy, short-wavelength radiation; lower-energy molecules give out lower-energy, longer-wavelength radiation.

The hang-glider gains height by turning to make sure that the wing stays in the warm convection currents of rising air. As the Sun beats down on land, some areas heat up faster than others. Newly plowed land, bare rock, or a beach reflect more heat than ground covered in vegetation, thus warming the air above them faster. This heated air becomes less dense and bubbles upward.

By harnessing the moving energy of these strong updrafts, or thermals, gliders and soaring birds can climb high into the sky. On a global scale, huge convection currents power the world's weather.

Sound: vibrant energy

Everyday sounds are caused by minute movements of the tiny molecules of the gases that make up the air around us.

Sound we hear is caused by vibration of the air. But vibration can be transmitted through any substance, whether solid, liquid, or gas, in which adjacent particles of the substance come into contact. Sound is, in fact, a pulse of energy transmitted through a substance by movement of the atoms or molecules of which it is made. Where a pulse passes, the particles in the substance are first squeezed together and then spring apart. It is like train cars in a siding. If a row of cars is rammed by a locomotive, the first one hits the next, which bangs into the next and so on. The energy from the impact passes down the line, but the cars stay more or less in the same place.

A single pulse can be started off by a single event – when hands are clapped and the air between them is forcibly compressed. Regular pulses can be started by an object that is vibrating, such as a stretched wire inside a piano. The number of pulses in the air matches the rate of vibration in the wire. The number of vibrations (or cycles) per second gives the frequency of the sound.

Sound travels fastest in solids such as steel, more slowly in water, and yet more slowly in air. At everyday pressures and temperatures, sound travels at about 760 mph (1,220 km/h) in air. In an electric storm, the light from the flash travels at about 186,400 miles per second (300,000 km/s), reaching an observer virtually instantaneously. Sound travels only 1 mile in 5 seconds (1 kilometer in 3 seconds). The distance of the flash can be worked out by counting the seconds between the flash and the sound. At some distance from the lightning-stroke, thunder does not crack, it only rumbles. This is because low-frequency sounds bend around obstacles such as hills and houses, whereas high-frequency sounds are more quickly absorbed than low-frequency sounds. Also, sound can be bent and reflected like light. The echo from a cliff face is exactly comparable to the image of an object in a mirror.

Because sound pulses set particles vibrating, as a sound passes, some of its energy is given to them. This means that sound energy is gradually dissipated as it moves through a substance and is converted to heat energy.

At a pop concert the audience both hears and makes sounds. The sound that is heard comes from loudspeakers which vibrate, causing vibrations in the air, which are then carried to the ears of the audience. Each hand clap from the audience in time to the music makes the air vibrate and creates sound.

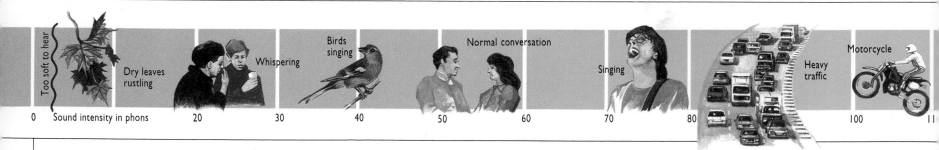

Too soft to hear | Dry leaves rustling | Whispering | Birds singing | Normal conversation | Singing | Heavy traffic | Motorcycle

0 — Sound intensity in phons — 20 — 30 — 40 — 50 — 60 — 70 — 80 — 100 — 11

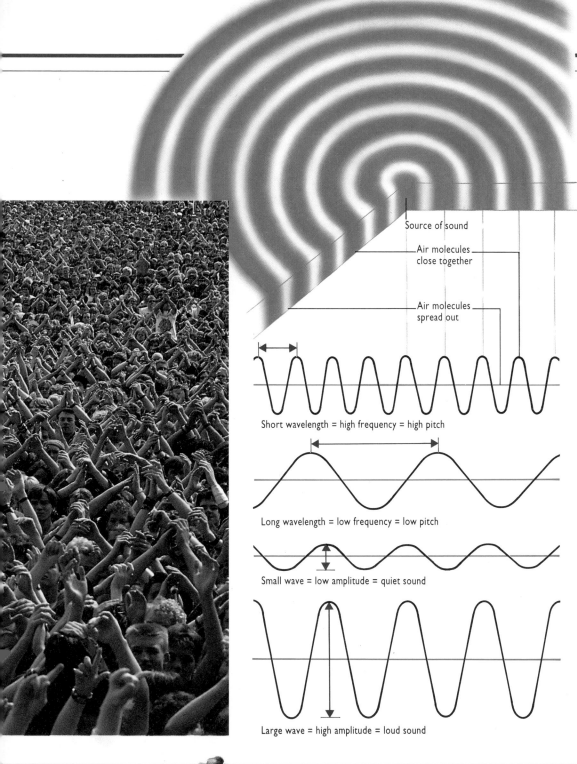

Source of sound

Air molecules
close together

Air molecules
spread out

Short wavelength = high frequency = high pitch

Long wavelength = low frequency = low pitch

Small wave = low amplitude = quiet sound

Large wave = high amplitude = loud sound

When transmitting sound pulses, air molecules oscillate back and forth. Movement of molecules drawn on a graph reflects the distance between them and therefore the pressure of the air as the pulse passes any given point. When sound is represented this way, it traces out a wave-shape.

The distance between the high-pressure points in the wave is the wavelength of the sound. The number of waves that passes a point each second is the frequency of the sound. The greater the movement of the air molecules, the more the pressure and thus the greater the amplitude (height) of the wave and the louder the sound.

The loudness of noise on the scale (**below**) is measured in phons, which are more accurate than decibels because phons take into account how we perceive sounds of different frequencies. Zero is the faintest, barely audible, noise; while values of 120 and more are damagingly loud and can cause injury.

The loudness of sound depends on the pressure of its pulses. Sound pulses of a conversation have about a hundred-thousandth of atmospheric pressure. When its pulses are about a thousandth of atmospheric pressure, a sound is painfully loud.

The scale compresses an enormous range of loudness. The power of a noise of 120 phons is a million million times more than that of the faintest sounds.

See also

► **ENERGY**
What is pressure?
54/55

► Energy's endless flow
58/59

► The energy spectrum
70/71

► Good vibrations
86/87

► Sound effects
88/89

► **ATOMS AND MATTER**
Gases: atoms set free
100/101

► **LIFE**
Sound signals
162/163

► **BRAINS AND COMPUTERS**
Sound sense
196/197

Human pain threshold

Rock band

Thunder overhead

Dangerously loud

Jet engine

Shot from rifle

Rocket launch

Nuclear explosion

120   130   140   150   160   170   180   190   200   210   220

Good vibrations

Music is in the ear of the beholder, but there are reasons why some sounds are harmonious and others just a noise.

Music is pulses or waves of vibration in the air made by musical instruments ranging from the violin to the human voice. Noise is a jumbled mix of many frequencies, but music has tones, or musical sounds, of a certain pitch, or dominant frequency. Middle C, for instance, has a frequency of 256 hertz. Musical sounds are not single frequencies, however, but complex mixes which give music its timbre, or "color."

Each instrument has its own timbre, which is why middle C on a violin and a clarinet sound different. While the main sound from each has a frequency of 256 hertz, each instrument also produces a unique range of additional higher frequencies (harmonics) which give it its timbre.

Whether or not notes sound pleasing together depends on harmony. Generally, harmonious notes are simple ratios of frequencies apart. For instance, any note is twice the frequency of the note an octave below it. In western music, other particularly harmonious notes include those with frequencies ³⁄₂, ⁴⁄₃, and ⁵⁄₄ times the frequency of a lower note.

Aerophones make music by causing a column of air to vibrate. The air is set vibrating in one of several ways: by blowing across a hole in a tube, by making a reed vibrate, or by the vibration of a player's lips against a mouthpiece. The pitch is set by the length of the vibrating air column – changing the tube's effective length changes the pitch. In many woodwind instruments, tube length is changed by exposing or covering holes in the tube. Another method is by opening or closing valves that send the air around lengths of tubing, as in brass instruments.

 Chordophones all make their sounds when a length of stretched string is made to vibrate. Strings are set vibrating by being plucked (harp), bowed (violin), struck (piano), or blown over (aeolian harp). Sound from each string is weak, so most chordophones have hollow, boxlike mounts that amplify the string's vibrations and add extra tones. The factors controlling the pitch of the tone produced are the length, tension, and weight per unit length of the string. The shorter, tighter, and lighter the string, the higher the pitch of the sound.

 Idiophones make sound when they themselves vibrate. Noises from this group include the crashes of cymbals, the clacks from castanets, and the "ting" of a triangle. Idiophones are played by being struck, as in the bells and gongs, plucked, as in the jew's-harp, banged together, as in cymbals, stamped on the ground or shaken, as in rattles and jingles. Most of these instruments do not produce musical tones, except for the bells, gongs, and xylophone-like instruments which can be used to carry a tune.

See also

ENERGY
▶ What is pressure? 54/55

▶ Sound: vibrant energy 84/85

▶ Sound effects 88/89

LIFE
▶ Sound signals 162/163

BRAINS AND COMPUTERS
▶ Sound sense 196/197

▶ Getting the message 202/203

 Membranophones make a noise when a membrane vibrates. Almost all are drums, and all must be hit to make them vibrate. Drums can be tuned to specific tones, as in the kettledrum, and the pitch depends on the size, weight, and tension of the membrane. A membranophone that is not drumlike is the mirliton, or kazoo, which makes a sound when vibrating air is blown past a membrane which is then set vibrating.

 The human voice is an extremely flexible and expressive instrument. Not only does it cover a wide range of notes, but the timbre can be changed. For instance, while singing a note a singer can make an "ah" sound and an "ooh" sound, as well as several others, by changing the volume of the mouth and shape of the lips. The voice works when the vocal cords – flexible flaps in the voice-box in the throat – are set vibrating when air from the lungs is forced past them. The tension in the chords controls the pitch.

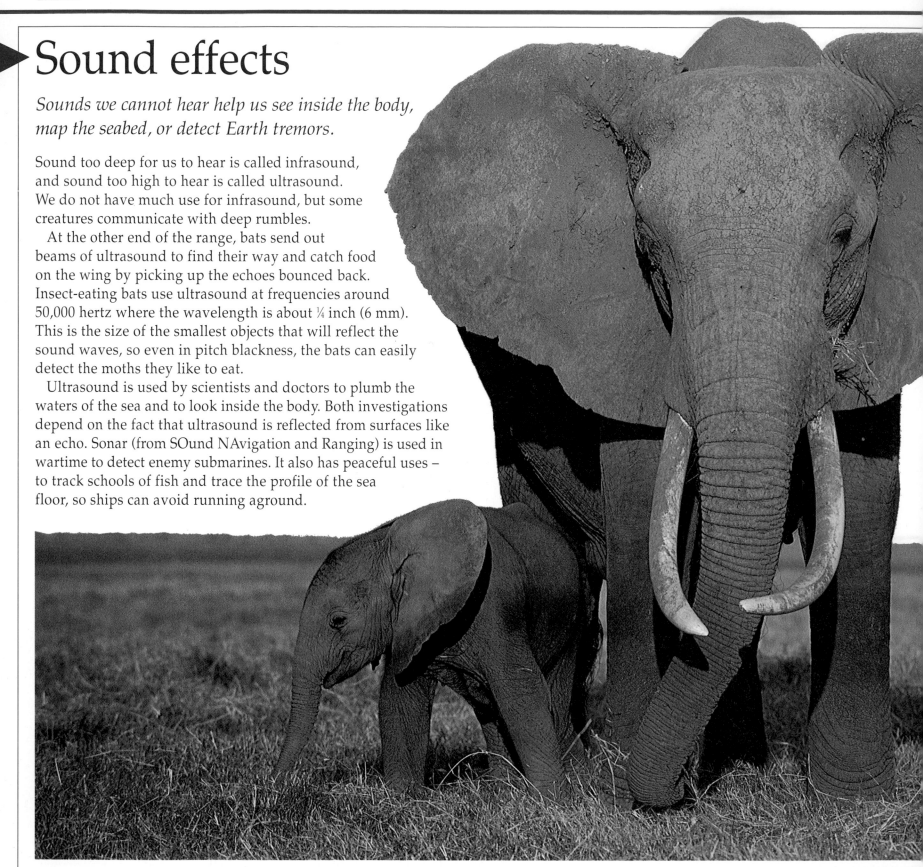

Sound effects

Sounds we cannot hear help us see inside the body, map the seabed, or detect Earth tremors.

Sound too deep for us to hear is called infrasound, and sound too high to hear is called ultrasound. We do not have much use for infrasound, but some creatures communicate with deep rumbles.

At the other end of the range, bats send out beams of ultrasound to find their way and catch food on the wing by picking up the echoes bounced back. Insect-eating bats use ultrasound at frequencies around 50,000 hertz where the wavelength is about ¼ inch (6 mm). This is the size of the smallest objects that will reflect the sound waves, so even in pitch blackness, the bats can easily detect the moths they like to eat.

Ultrasound is used by scientists and doctors to plumb the waters of the sea and to look inside the body. Both investigations depend on the fact that ultrasound is reflected from surfaces like an echo. Sonar (from SOund NAvigation and Ranging) is used in wartime to detect enemy submarines. It also has peaceful uses – to track schools of fish and trace the profile of the sea floor, so ships can avoid running aground.

Some sounds, or vibrations, pass right through the Earth. These are shock waves sent out by earthquakes and other earth tremors and are detected by a device called a seismograph. Mapping waves traveling through the Earth shows that Earth's density rises at the center and that it has a solid inner and a liquid outer core.

An ultrasound scanner uses high-frequency sound – at around 10 million hertz – to show a baby in the womb. Ultrasound is reflected at the boundary between tissues, so the screen image is made from echoes from inside the body.

See also

▶ **ENERGY**
Energy's endless flow
58/59

▶ Sound: vibrant energy
84/85

▶ Good vibrations
86/87

▶ **SPACE**
Land, sea, and air
14/15

▶ **LIFE**
Sound signals
162/163

▶ **BRAINS AND COMPUTERS**
Sound sense
196/197

Elephants communicate with subsonic rumbles too deep for us to hear. The deep sounds they make – so low-pitched the air seems to throb – travel large distances, unlike higher-pitched sounds, which are absorbed by vegetation. Using these sounds a female elephant in heat can attract a mate, and groups of elephants can stay in touch despite being out of sight of each other.

SHIFTING SOUND

The change in pitch of a train horn as the train races past is due to the Doppler effect. As the train nears a listener, each successive sound wave has slightly less far to travel than the one before it. As a result, there is less time between the arrival of one sound wave and the arrival of the next, and the pitch of the sound is raised. When the train moves away, each wave has farther to travel than the one before it, so fewer arrive each second, and the pitch is lowered.

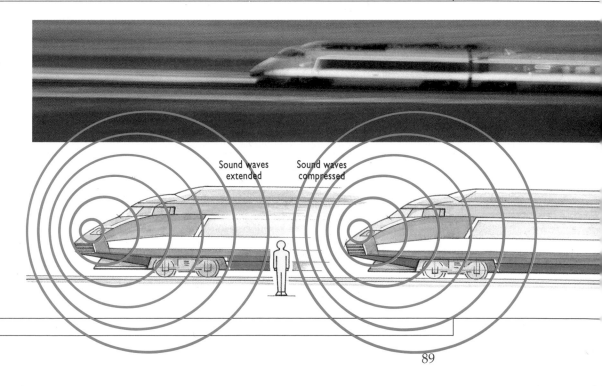

Sound waves extended

Sound waves compressed

Falling for gravity

There is no escape from the effects of gravity; it acts over the vastest distances imaginable. Closer to home, it gives us weight and tells us which way is "down."

It is gravity that gives objects weight. In fact, on Earth the weight of an object is actually the downward gravitational force which is acting upon it. This force is directly related to the object's mass, or in other words, to the amount of matter it contains. For instance, when a bag of flour is weighed, what is really being measured is the force due to gravity that the flour exerts on the weighing machine, and this tells us how much flour there is in the bag.

Gravity is, in fact, one of the four fundamental forces in the universe, and it acts between any two objects, making them attract or move toward each other. This means that every object in the universe, however small or large, that has mass is attracting every other object that has mass.

The attractive force of any object makes other objects move toward it with ever-increasing speed – that is, it makes them accelerate. Near the surface of the Earth, falling objects pick up an extra 32 feet per second (9.8 m/s) of speed for each second that they fall. This is the acceleration

WEIGHTS ON DIFFERENT WORLDS

The gravity on a planet's surface is related both to the planet's mass and to the distance of the surface from the center of the planet's mass. So although the Moon is only 1/80 of the Earth's mass, the gravity at the Moon's surface is much more than 1/80 the gravity at the Earth's surface. This is because the Moon's surface is about four times closer to its center of mass. The net result is that the gravitational force felt on the Moon is only 1/6 of that felt on Earth's surface so objects weigh just 1/6 of their weight on Earth.

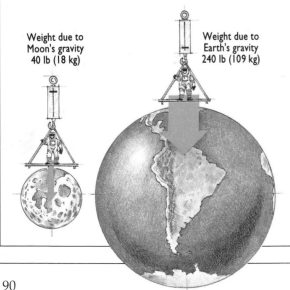

Weight due to Moon's gravity
40 lb (18 kg)

Weight due to Earth's gravity
240 lb (109 kg)

An Apollo astronaut on the Moon's surface is not weighed down by his spacesuit, with its thick layers, oxygen tanks, and radio, because the suit weighs only 1/6 what it would weigh on Earth. For the same effort, he could jump six times as high on the Moon as on Earth.

due to gravity at the Earth's surface; elsewhere in the universe, different gravitational forces produce different accelerations.

Gravity works over vast distances, but the attraction falls off in proportion to the square of the distance between objects. If you double your distance from the center of mass of a large object (like a planet), you feel only one-quarter the gravitational force. Triple the distance, and you feel only one-ninth the force of gravity.

Gravity is the weakest of the fundamental forces. It is so weak that the whole bulk of the Earth, pulling down on a pin, can be overcome by a toy magnet attracting the pin and pulling it up. But because gravity works over such long distances, gravity's infinite range dominates the universe.

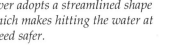

When sky divers step *out of an aircraft (**above**), they accelerate toward the Earth's center of mass under the influence of the Earth's gravitational force. But they do not continue to pick up speed all the way down to the Earth's surface.*

As their speed increases, the friction between them and the air also increases and their rate of acceleration falls. When they reach a certain speed, the air resistance matches the force due to gravity and the sky divers stop accelerating and stay at a speed called the terminal velocity. When the parachute is opened the air resistance increases and the sky divers slow down enough to land safely.

In theory, the mass of each sky diver attracts the Earth also, but because this mass is minute compared with that of the Earth, there is no noticeable effect.

The force of Earth's gravity *accelerates the diver (**left**) toward the center of the Earth's mass. The diver adopts a streamlined shape which makes hitting the water at speed safer.*

A prize fish is registered by weight on a spring-scale. The scale measures weight by showing how much a spring stretches when the fish is hung from it. The spring resists more and more as it stretches until its upward tension equals the downward weight due to the force of gravity.

Defying gravity

The planets and stars keep their orbits, meteors hurtle through space and astronauts become weightless by defying gravity. How?

Gravity pulls together everything from the tiniest subatomic particle to the biggest galaxy. Yet there are times when it is impossible to feel gravity, even close to objects as big as stars. Things become weightless; that is, they cease to feel the force of gravity. Astronauts in space, for instance, appear to be weightless as they float around the cabins or drift outside on a space walk. But there are other, less familiar, examples of things defying gravity. When it is thrown, a ball flies through the air because it is momentarily weightless. And if you weigh yourself while descending in a high-speed lift, you might find you miraculously shed a few pounds as the lift gathers speed.

Gravity accelerates things toward each other. If a gravity force is matched by another acceleration in a different direction – or if everything experiences the same acceleration – the effect of gravity may be reduced or even appear to be cancelled out. When an astronaut is in space, the pull of the Earth's gravity is only marginally less than on the ground, so he has not really lost "weight." It is just that the acceleration due to gravity is balanced out by another acceleration. The astronaut is, in fact, falling toward the Earth with the acceleration due to gravity. But because he is also speeding around the Earth, in the time it takes to fall down a certain distance, he has moved far enough around the curve of the Earth to still be the same distance above it. So although he is falling toward Earth, he never gets any closer to it.

This effect keeps the Moon circling the Earth, the Earth and planets circling the Sun, and all the stars in each galaxy circling around each other in a giant cosmic dance. Planets and stars keep orbiting each other at the same distance rather than being pulled slowly together because they are traveling through space at high speeds – speeds fast enough to balance exactly the gravitational attraction. If the speed were too high, they would fly off into space; if it were too low, they would gradually fall together.

An astronaut floating in space is forever falling toward the ground, but he is hurtling around the Earth so fast he never gets any closer. So, too, are the spacecraft and all its contents, which is why they appear weightless. They are all free-falling together. The space traveler cannot feel the pull of gravity or his enormous speed around the Earth, but he can see the Earth speeding past beneath him all the time.

Prolonged exposure to weightlessness can have strange effects on a body used to coping with gravity. Blood rushes to the head, puffing up an astronaut's face; and, on very long trips, the bones lose calcium and become progressively weaker.

Artificial satellites stay orbiting the Earth because they are speeding around the world just fast enough to stop them from falling. To maintain the correct orbit, therefore, satellites must be launched into orbit at exactly the right height and speed. The critical speed is 17,700 mph (28,500 km/h). If the satellite orbited any slower, it would fall back to Earth. At the correct speed, it goes into a circular orbit about 100 miles (160 km) up, just above the atmosphere. If it reached the Earth's escape speed of 25,200 mph (40,554 km/h), it would break loose from the effects of Earth's gravity and not return.

Interplanetary probe launched with sufficient velocity to escape Earth's gravity

Low-velocity weather rocket falls to ground

Satellite moving at orbital velocity "falls" around Earth in circular orbit

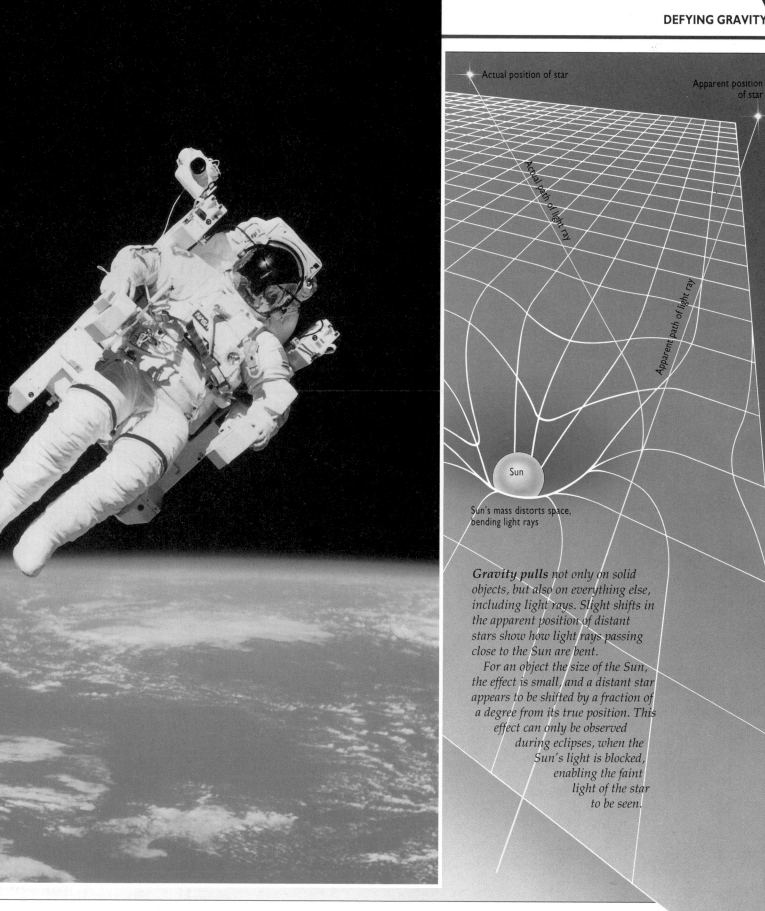

Actual position of star

Apparent position of star

Actual path of light ray

Apparent path of light ray

Sun

Sun's mass distorts space,
bending light rays

Gravity pulls not only on solid
objects, but also on everything else,
including light rays. Slight shifts in
the apparent position of distant
stars show how light rays passing
close to the Sun are bent.

For an object the size of the Sun,
the effect is small, and a distant star
appears to be shifted by a fraction of
a degree from its true position. This
effect can only be observed
during eclipses, when the
Sun's light is blocked,
enabling the faint
light of the star
to be seen.

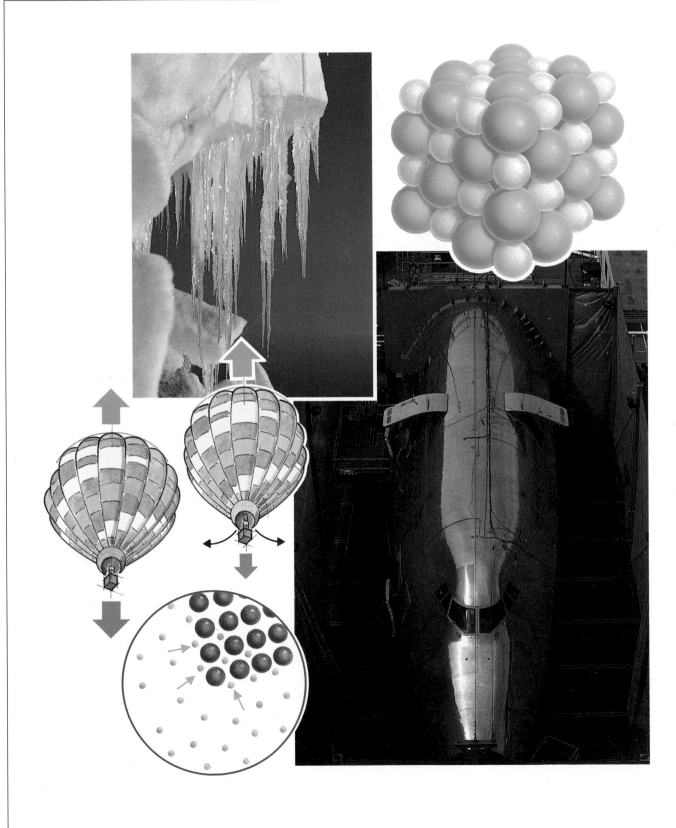

Atoms and Matter

Everything in the universe is made of matter, from the thinnest wisp of gas to the densest planet. Matter has three fundamental forms: solid, liquid, and gas, and is made up of elements, mixtures, and compounds. But a simple understanding of the kinds of matter is enormously enhanced by an in-depth knowledge of the atom – the smallest chunk of matter for most practical purposes.

An understanding of the structure of the atom and how atoms interact provides profound insights into the substances that we see around us and into the way they change, with examples from plastic shoes to burning candles.

But atoms are not the smallest chunks of matter; they are made from particles so tiny that each atom is largely empty space. The more we learn about matter, the more insubstantial it seems to become. Even tiny subatomic particles are not tangible solid balls, but little packets of energy held together by curious forces.

Left clockwise from top: icicles, matter changing form; atoms in a crystal structure; metals mixed in alloys; solutions; gases expanding. **This page (top):** Water of life; (**left**) atoms splitting to make energy.

Solid structures

Solids are characterized by the fact that their atoms and molecules are arranged in closely packed structures.

Rock is solid, water is liquid, and air is gas; but they do not have to be. Under the right conditions, every substance can be either solid, liquid, or gas. Even rock melts and then boils away if the temperature is high enough. Similarly, in the laboratory, all the gases in air can be frozen solid by reducing the temperature to about 3K: -454°F (-270°C).

The states of matter – solid, liquid, and gas – are a reflection of the three fundamentally different ways the particles (atoms and molecules) in a substance can pack together.

In a solid, the atoms are packed very closely – so close, in fact, that they cannot move about freely as they do in gases and, to a lesser extent, liquids; they can only vibrate and spin on the spot. The usual effect of this close packing not only makes solids much denser than either liquids or gases, but also gives them a definite shape and volume. It is sometimes said solids have a fixed shape, while liquid and gases flow out to fill the container they occupy. Although this is generally true, there are many exceptions. Solid ice in a glacier, for instance, can flow without actually melting, just as tar on a road can creep and flow on a hot day.

In nearly all solids, the particles are set in a regular pattern, and this means that they are a form of crystal, the shape of which depends on the size and shape of the particles and the strength of the bonds between them. When crystals are heated, they melt to liquid at a particular temperature, called the melting point. The weaker the bonds between the particles, the lower the melting point.

There are some solids, such as glass and some resins, which do not have a crystal structure, but are amorphous – have no set shape. Amorphous solids have no fixed melting point, but simply become more pliable as they get warmer.

Most solids are made up of crystals. These crystals (**left**) are all types of the mineral beryl which is a substance made mainly of the elements oxygen and silicon with beryllium and aluminum as well. Crystals of beryl always have a definite shape which is determined by the way the atoms of the different elements fit together. The most common form of beryl is hexagonal (six-sided). Pale blue beryl is called aquamarine; dark green beryl is the precious gem emerald. Other forms of beryl are heliodor (yellow) and morganite (rose-colored).

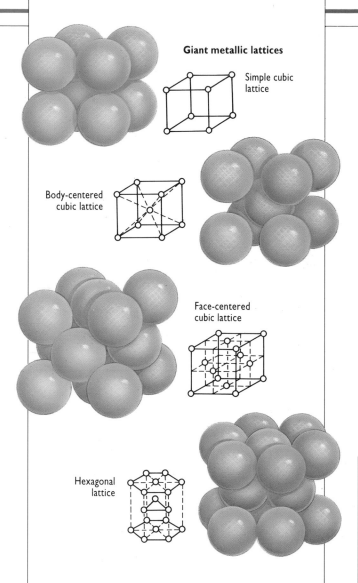

Giant metallic lattices

Simple cubic lattice

Body-centered cubic lattice

Face-centered cubic lattice

Hexagonal lattice

*The atoms in most crystals are packed as close as possible. How they pack depends on how the atoms bond to each other. Two common arrangements are the cubic form and the hexagonal form (**above**).*

Copper, a metal, forms simple cubic crystals in which the atoms are arranged like the corners of a simple square, and the crystals have six sides. In the body-centered cubic structure, typified by crystals of iron, atoms are more closely packed and are generally set in a cube around a single atom.

More complicated cubic crystals, such as the face-centered cubic crystals of common salt (sodium chloride), have extra atoms sitting in the faces of each side of the square. The atoms are arranged in layers which adjoin in such a way that the atoms form a cube every other layer. Hexagonal crystals, such as those formed by zinc, have six sides, and the atoms are arranged in layers.

In a solid, the strong bonds between molecules and atoms means they keep their shape – which is why they can be used for building. For instance, the Great Pyramid of the Pharaoh Khufu at Ghiza in Egypt has kept its near-perfect symmetry for almost 3,500 years. The pyramid is built from 2.3 million blocks, weighing on average 2.5 tons each, of granite – an extremely tough and long-lasting building material. Granite, a rock formed during volcanic activity, is made mainly from crystals of three minerals: quartz, mica, and feldspar. All of these are silicates, consisting of regular structures of silicon and oxygen atoms. In quartz the atoms are arranged in pyramids; in mica, they are set in sheets.

Solids can, of course, be broken. Structural failures occur when the bonds between the particles in a solid are weakened. In crumbly substances, such as cookies, this happens easily. In metals such as steel, bonds are much stronger. Despite the strength of metals, repeated stress in the same place can weaken bonds enough to make them liable to sudden failure. This is called metal fatigue. Metal aircraft bodies are frequently X-rayed for any signs of weakness, especially in safety-critical components.

See also

ATOMS AND MATTER
▶ Molecules in motion 98/99

▶ Gases: atoms set free 100/101

▶ Metal connections 114/115

▶ Water: the unique molecule 124/125

ENERGY
▶ Beyond violet 76/77

▶ A matter of degree 80/81

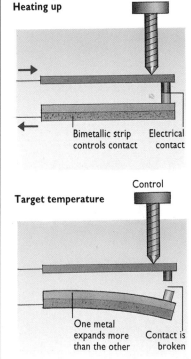

Heating up

Bimetallic strip controls contact — Electrical contact

Target temperature

Control

One metal expands more than the other — Contact is broken

HEATING AND BENDING

An iron's thermostat exploits the way substances expand when heated. When a solid absorbs heat energy, the molecules vibrate more, move a little farther apart, and the substance expands. Just how much farther depends on the substance.

The thermostat in an iron uses a strip of two metals with different coefficients of expansion – that is, metals that expand at different rates. The two metal alloys in this bimetallic strip – invar and brass – are welded together.

Brass expands more than invar, so the strip bends when it gets hot. When the iron gets hotter than it should be, the strip bends and breaks an electrical contact, switching off the iron's heating coil so it cools.

Molecules in motion

In a liquid, particles are held only loosely together, allowing it to flow in all directions.

When a solid melts to a liquid, it changes dramatically. The heat needed for melting makes particles vibrate so much that the bonds between them are weakened. In a solid, particles are locked in a regular pattern, but in a liquid they move significantly farther apart – which is why a liquid takes up, on average, at least 10 percent more space than a solid.

Bonds between particles in a liquid are so weak they cannot hold the liquid in a definite shape. So liquids are fluid, which means they can usually be poured and that they flow, under the influence of gravity, to fill the bottom of any container they are in, leaving a smooth, level upper surface. Just how quickly they flow depends on their viscosity or thickness. In a thick, highly viscous liquid, like molasses or cold engine oil on a winter's morning, the bonds between particles are strong enough for the flow to be slow; thinner, less viscous liquids, such as water, flow much more rapidly.

The bonds between particles in a liquid are actually so loose that particles not only vibrate and spin on the spot as they do in a solid, but often break away to wander around freely. This is why liquids mix easily, as particles diffuse from one part of the liquid to another.

The idea that particles in a liquid are

MELTING MOMENTS

As a solid is heated, its particles vibrate more and more until the bonds between them can no longer hold and the solid disintegrates. As a solid melts, its temperature stays at the melting point while any of the solid remains. When ice melts, the ice stays at 32°F (0°C) until it has all melted.

Substances such as ice, in which particles are held loosely together, melt at low temperatures. Where atoms are held in a strong array, as in diamond, the substance is solid even at high temperatures. If substances have a similar structure, the heavier the atoms, the higher the melting point. The reason is that heavy atoms need more energy to loosen bonds.

Melting

Vibration of particles increases

Increasing heat

Solid

In a liquid, bonds between particles are so weak that it flows freely. So when rainwater falls on the landscape, some soaks into the ground – either bubbling out from springs farther down, or sitting underground as permanent groundwater. The rest runs downhill under the influence of gravity into the lowest points of the landscape, filling up river channels and creating lakes, some of which are intricately shaped (left).

constantly moving is borne out both by the evidence of diffusion and by a phenomenon first spotted by British botanist Robert Brown in the 1820s. Brown observed through his microscope that pollen grains suspended in water jittered about in a random manner. This jittering, called Brownian motion, is caused by the constant jostling of the pollen by moving water particles.

Since particles in a liquid are forever on the move, some may have enough energy to escape their bonds altogether, thus helping the liquid evaporate to a gas. And the higher the temperature of the liquid, the more particles have the energy to escape. Because the particles that escape are those with more energy than the rest, evaporation makes the liquid cooler because only low-energy particles are left behind – which is why sweating cools you down when you are hot and why damp clothing feels cold.

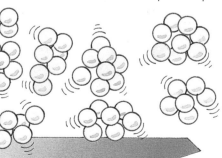

Lattice breaks up into clumps

Liquid

When a solid melts, the vibration of warming particles becomes so great that bonds in the regular lattice can no longer hold them and clumps of particles break away (left). When a liquid freezes (right), the particles in it vibrate less energetically as the temperature drops. Soon, their mutual attraction is enough to draw them together, first into clumps here and there, but eventually into a tightly knit structure as the substance becomes solid once more.

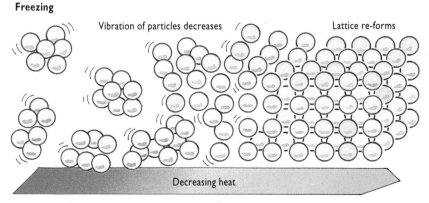

Freezing

Vibration of particles decreases

Lattice re-forms

Decreasing heat

Liquid

Solid

Gases: atoms set free

The very air we breathe is, like other gases, made up of rapidly moving tiny particles.

The thinnest possible form of matter, gases consist largely of empty space. Even in the densest gas, just a minute fraction of the volume is taken up by gas particles. When a gas condenses to a liquid, the liquid takes up, on average, 1,300 times less space. Because there are so few particles in any given volume, gases are usually transparent.

Particles in a gas are so energetic and widely spaced that any attractions between them are not strong enough to hold them together. So they zoom around all over the place, crashing into each other billions of times a second – which is why gases usually spread out as far as they can to fill up any container completely. Just how fast the particles move depends on the temperature; the higher the temperature, the faster they move. At room temperature, oxygen particles in air hurtle around at over 1,000 mph (1,600 km/h) on average, though the speeds of individual particles vary widely.

These high-speed particles constantly bombard anything in their path, and so gases exert continuous pressure on surrounding surfaces. The more particles there are hitting a given area, the greater the pressure, so if a gas is squeezed into a smaller space, its pressure rises. For instance, when a bicycle tire is pumped up, the air is forced inside under pressure.

Evidence for the speed of movement of gas particles comes from the fact that gases rapidly mix together, a process called diffusion. For instance, if you open a bottle of strong perfume in a small room, its scent can soon be detected throughout the room. For this to happen, the scent particles obviously have to move rapidly, mixing in with the surrounding air particles until they are evenly distributed. Similarly, scent from an incense burner spreads rapidly through a church as the burner releases scent particles. How fast two gases mix depends on how dense they are. The denser a gas, the more slowly it diffuses. In 1829, British scientist Thomas Graham showed that the relative rates at which gases diffuse are inversely proportional to the square roots of their densities. This is called Graham's Law.

Hot-air balloons take off because heat from the burner makes the particles in the air bag, or envelope, move faster. They collide with each other harder and spread out, inflating the envelope and making the air inside lighter than the air outside. Lighter air in the envelope gives the balloon upthrust, which lifts the balloon. But the atmosphere is thinner higher up, so the number of particles bombarding the outside lessens, reducing pressure and allowing air inside to expand.

Since the air in the balloon expands without gaining energy, the same amount of heat is spread through a bigger space, making it cooler. As the air becomes cooler, the balloon begins to sink again. To maintain height or to rise, the balloonist has to fire the burner to re-heat the air.

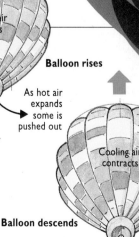

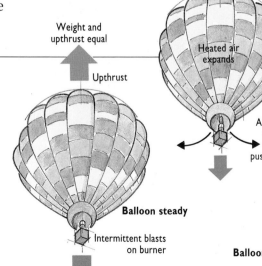

Weight and upthrust equal

Upthrust

Heated air expands

Balloon rises

As hot air expands some is pushed out

Balloon steady

Intermittent blasts on burner

Cooling air contracts

Balloon descends

Weight

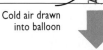

Cold air drawn into balloon

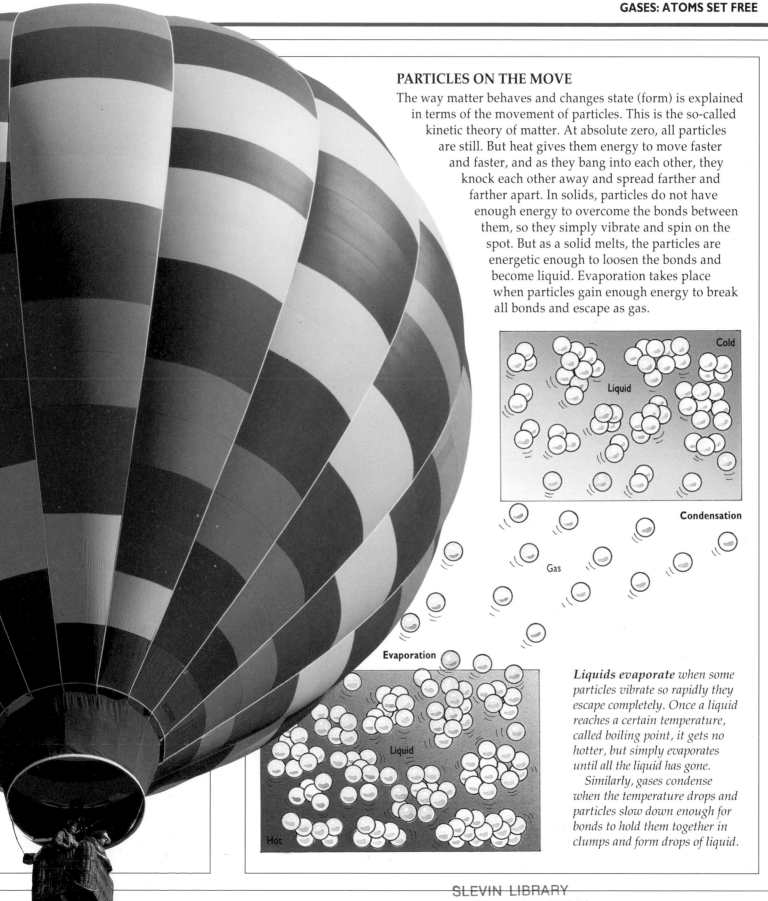

PARTICLES ON THE MOVE

The way matter behaves and changes state (form) is explained in terms of the movement of particles. This is the so-called kinetic theory of matter. At absolute zero, all particles are still. But heat gives them energy to move faster and faster, and as they bang into each other, they knock each other away and spread farther and farther apart. In solids, particles do not have enough energy to overcome the bonds between them, so they simply vibrate and spin on the spot. But as a solid melts, the particles are energetic enough to loosen the bonds and become liquid. Evaporation takes place when particles gain enough energy to break all bonds and escape as gas.

Cold

Liquid

Condensation

Gas

Evaporation

Liquid

Hot

Liquids evaporate when some particles vibrate so rapidly they escape completely. Once a liquid reaches a certain temperature, called boiling point, it gets no hotter, but simply evaporates until all the liquid has gone.

Similarly, gases condense when the temperature drops and particles slow down enough for bonds to hold them together in clumps and form drops of liquid.

Matters of substance

Every one of the vast range of substances in the universe is either an element, a compound, or a mixture.

The world contains such a huge variety of substances it is hard to imagine what they have in common. Yet all are made from barely more than 100 basic chemicals, called elements, such as gold, oxygen, and carbon, each with its own unique atom and character. All the atoms in a pure sample of an element such as pure carbon are identical – but different from the atoms of every other element.

What gives an atom of an element its own special character is simply the number of protons in its nucleus, which is usually matched by the number of electrons flying around the nucleus. Hydrogen has one proton, helium two, carbon six, and so on. The maximum number of protons found to date in a nucleus is 109, but there are actually only 106 known elements – this is because atoms with 106, 107, or 108 protons have yet to be found. Gold is an element with 79 protons in its nucleus, which is why gold has an atomic number of 79. There are also 118 neutrons, making gold atoms big compared with most others. The combination of 118 neutrons and 79 protons gives gold atoms a mass number of 197, compared with 27 for aluminum and just 1 for hydrogen, the tiniest atom of all.

The atoms of some elements can exist separately, but they are more usually linked together in clusters of two or more, called molecules. These are the smallest part of an element that can normally exist alone. Chlorine, for instance, usually consists of molecules made from pairs of chlorine atoms. Only at very high temperatures can chlorine atoms exist singly.

Element

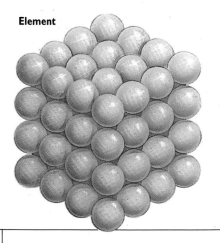

*The mask of the ancient Egyptian Pharaoh Tutankhamun is made of gold, a metallic element. In gold the identical atoms of the element are packed closely together (**left**). Each element has its own unique properties. Gold has long been prized because it is relatively soft and easy to work with, it has an attractive luster, and it does not react easily with other elements to make compounds with different properties.*

Iron, another useful element, is harder than gold, but is prone to combine with oxygen in the air to form rust, a crumbly compound much weaker than iron itself.

Some molecules are made from two or more different atoms linked together. These are called compounds. Common salt is a compound of sodium atoms and chlorine atoms. There is an enormous range of compounds, ranging from those with just two different atoms in them, such as common salt, to those with millions of atoms in them, such as DNA. Pure elements and compounds are rare, however, and most substances are mixtures, containing a variety of different compounds and elements mixed together. Food, milk, air, and sea water are all mixtures. Pure water is a compound made of hydrogen and oxygen atoms, but most water has other compounds, such as calcium bicarbonate, dissolved in it.

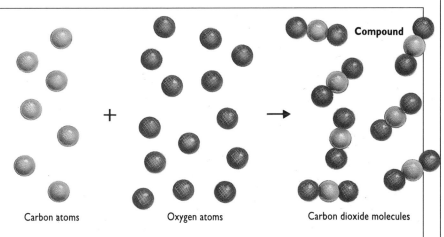

Mixture

Water molecule

Ethanol molecule

Maltose molecule

Like most substances, whiskey (**above**) is a mixture, made of about 40 percent ethanol (alcohol) and 60 percent water, with traces of other compounds, such as maltose, for flavor. The different compounds in a mixture are not chemically joined, so they can often be physically separated. In whiskey production, the ethanol is extracted by distillation (**left**), which makes use of the fact that water and ethanol have different boiling points. The mixture is heated to a temperature where just the ethanol boils. Ethanol vapor condenses at the top of the still, where pure ethanol is drained off to be mixed with water in the right proportions.

One commonly used compound is carbon dioxide. It forms the bubbles of gas found in a glass of sparkling mineral water. Carbon dioxide is a compound of carbon and oxygen, and every single molecule is made of one carbon atom and two oxygen atoms. Like elements, chemical compounds have their own unique properties. For instance, carbon dioxide always remains solid as long as it is kept under pressure or below -108°F (-78°C). But as soon as it is exposed to normal room temperatures, it changes to carbon dioxide gas, rather than melting to a liquid.

Carbon atoms

+

Oxygen atoms

→

Compound

Carbon dioxide molecules

Inside the atom

If you could take a look inside the atom, you would probably see nothing – atoms are mostly empty space.

Break down matter into smaller and smaller pieces, and eventually you come to the atom, the smallest possible particle of any substance. Atoms are so tiny that about six million of them could fit on the period at the end of this sentence.

Atoms were once thought to be the ultimate, unsplittable building blocks of the universe. The word "atom" comes from the Greek *atomos*, meaning indivisible, and until New Zealand physicist Ernest Rutherford smashed atoms of nitrogen in 1919 at the Cavendish Laboratory, Cambridge, England, it was thought impossible to split matter into anything smaller than an atom. Atoms were imagined to be like solid indestructible balls. In fact, scientists now think of them as being more like clouds of energy – mostly empty space dotted with even tinier subatomic particles.

The model of the atom accepted in the 1990s has a dense nucleus with electrons orbiting around it in various layers, called shells. In the nucleus of all atoms except hydrogen, there are both protons and neutrons.

The nucleus is actually minute compared with the space occupied by the atom. In 1911, Rutherford worked out, from what happened to a stream of particles fired at metal foil, that the nucleus was about 10^{-13} inch (10^{-14} m) across – just ten-thousandths of the size of the atom. If the whole atom were the size of a football stadium, the nucleus would be no bigger than a pea placed in the center of the field.

*Protons and neutrons are about 10^{-13} inch (10^{-15} m) across and have a mass of about 13.7×10^{-27} pound (7×10^{-27} kg). Electrons are much smaller and have a mass only about 1/1,836th that of a proton. If the nucleus was the size of a soccer ball, the electron would be the size of a pinhead lost in the middle (**far right**).*

One simple and obvious result of atomic structure is static electric effects, like the tingle you sometimes get when you take off an acrylic sweater and the way dry hair can stand on end when combed. Because the outermost electrons are far from the nucleus, they are held in place only weakly and are apt to get knocked off. As you drag the sweater over your head, some electrons are knocked off atoms in the sweater and are drawn to the other surface. One surface has more negatively charged electrons than the other. The two surfaces are drawn together by the opposite charge, and the charge is balanced out by a spark, down which electrons flow.

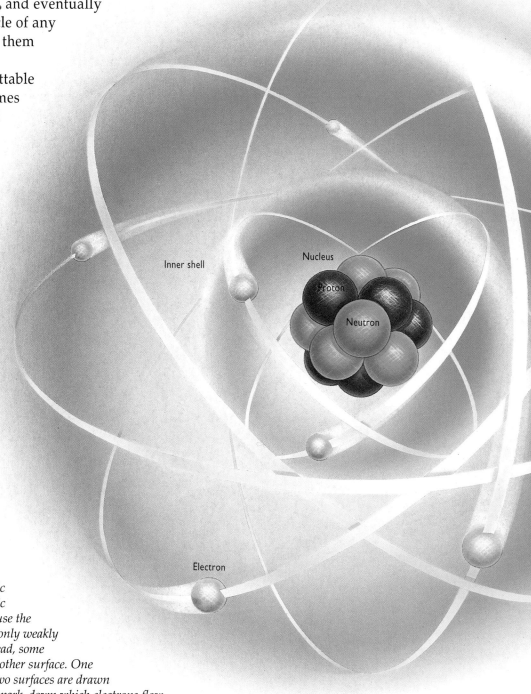

Outer shell

Inner shell

Nucleus

Proton

Neutron

Electron

Experiments have shown that there are more than 70 different subatomic particles. Only three, though, have any significant effect on the way materials behave, so chemists work with a model of the atom made of these three.

At the atom's heart is a dense nucleus made of two kinds of particles, protons and neutrons. Protons have a positive electrical charge; neutrons have none. Whizzing around the nucleus are much smaller negatively charged particles called electrons. Most atoms have identical numbers of protons and electrons, so the electrical charges balance each other, making atoms electrically neutral.

Atoms can be split, but they are usually bound together by three forces. Negatively charged electrons are held in orbit around the nucleus by their electromagnetic attraction to the positively charged protons. Protons and neutrons are bound together in the nucleus by strong and weak nuclear forces. With gravity, these three forces make up the four forces of the universe.

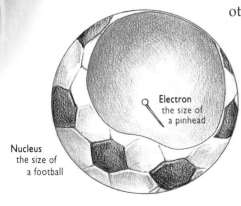

Nucleus
the size of
a football

Electron
the size of
a pinhead

Atoms can actually be seen. The best pictures of atoms are taken by scanning tunneling microscopes (STMs), developed by IBM in Zurich, Switzerland, in 1981.

This STM image (above) shows gold atoms deposited on a base of carbon atoms. The gold – shown as yellow, red, and brown – forms a layer just three atoms thick on top of the carbon atoms which are in the form of a layer of graphite.

STMs work by exploiting the difference in electrical charge between a special needle and the atom surface to create a picture of the atom like a contour map. The electrical difference causes a current to flow, but the strength of the current varies according to the position of the atom. So as the needle scans back and forth over the atom's surface, the microscope records these variations to create a map of the positions of the atoms.

The key to behavior

The nucleus, and the arrangement of the electrons around it, are the crucial components that dictate how an element will behave.

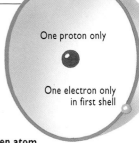

Hydrogen atom

If we could take millions of photographs of electrons zooming around the nucleus of an atom, they would appear each time in a slightly different place. The different positions form a series of up to seven rings of clouds or "shells" around the nucleus, where the chances of finding an electron are high. In the smallest atoms, hydrogen and helium, there is just one small shell close to the nucleus. Helium atoms have two electrons and hydrogen one, so the chance of finding an electron at a given point in this shell is twice as high with a helium atom as with hydrogen.

Carbon atom

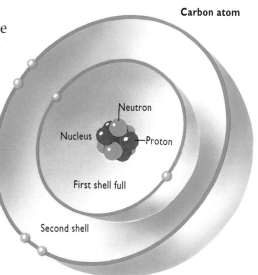

There is always a limit to the number of electrons each shell can hold. In the inner shell, there is room for only two, so if the atom has any more electrons, they move out to a second shell, farther from the nucleus. This second shell can hold up to eight electrons. The third shell, too, can take up to eight, but may hold more – up to eighteen – if there is another shell. The outer shell only exceptionally has more than eight electrons.

Atoms with eight electrons in their outer shell are very stable and slow to react with other elements. This is because it takes a great deal of energy either to add an extra electron or take one away. Atoms with one electron in their outer shell, such as hydrogen, sodium, and potassium, are very reactive because this electron is easily removed. Similarly, atoms which lack one of the eight, such as fluorine, chlorine, iodine, and bromine, called the halogens, are reactive because they readily accept another electron into the gap in their outer ring. Fluoride (a fluorine atom with an electron gained from another atom) in toothpaste protects teeth by forcing out and replacing a component of tooth enamel that is attacked by acids in food.

*As the space shuttle lifts off from its launch pad (**right**), the chemical reaction that powers its main engines is made possible by the atomic structure of two elements: oxygen and hydrogen.*

Two atoms of hydrogen, each with one electron in its outer shell, combine with one oxygen atom, with six electrons in its outer shell. The heat released expands the resultant compound – water – and sends it hurtling out of the base of the engines as fast-moving steam.

The shuttle uses an external fuel tank to hold the oxygen and hydrogen. As well as its main engines, the shuttle has two solid fuel booster rockets. These are discarded, along with the fuel tank, before the shuttle goes into orbit.

One electron only
in third shell

Second shell full

First shell full

Sodium atom

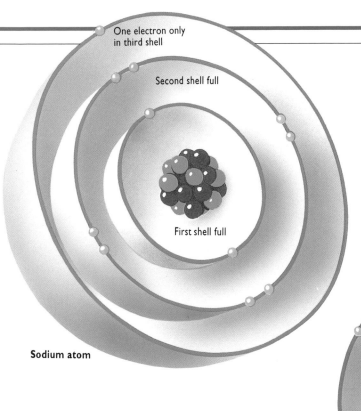

Hydrogen is the smallest and simplest of all atoms and the most abundant in the universe. It has just one proton in its nucleus, making it extremely light. It also has just one electron in its single electron shell. This makes it likely to react with other atoms.

Carbon has six electrons altogether, two in its inner shell and four in its outer shell. With four "vacancies" in its outer shell, it can form complex links with other atoms such as oxygen atoms, which have eight electrons altogether: two in the inner shell and six in the outer shell. Oxygen also readily shares electrons with two hydrogen atoms, filling up their outer shells and making water.

Sodium atoms have a total of 11 electrons: two in the first shell, eight in the second, and just one in the third, which is readily lost. In common salt, which is chemically known as sodium chloride, sodium gives this lone electron to a chlorine atom, which has a vacancy in its outer shell.

See also

ATOMS AND MATTER
▶ Inside the atom 104/105

▶ Classifying the elements 108/109

▶ From atoms to molecules 110/111

▶ The intimate bonds 112/113

▶ Metal connections 114/115

▶ Attraction of opposites 116/117

▶ Light from atoms 138/139

Second shell

First shell full

Oxygen atom

Helium atom

Shell full

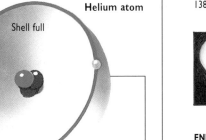

ENERGY
▶ Sparks and attractions 66/67

▶ The energy spectrum 70/71

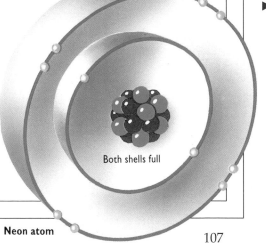

LIGHT AND SAFE

Airships are filled with helium gas, which is lighter than air because its atoms are small and light. Airships were once filled with hydrogen gas, which is even lighter. But because they have just one electron in their outer shell, hydrogen atoms are highly reactive, and the gas can catch fire. The element helium (**right above**), with a

full complement of two electrons in its single shell, is stable and safe.

An atom of neon (**right**) also has a full outer shell. Helium, neon, and the other elements which have full outer shells make up a group of gases that behave similarly. They are called the inert (or noble) gases because they are stable and unlikely to react with other elements.

Both shells full

Neon atom

Classifying the elements

If a list of the chemical elements is arranged in a certain way, definite – and useful – trends emerge.

Each of the elements has its own unique character, but certain patterns begin to emerge if they are arranged in order of the size of their atoms. This is the basis of one of the chemist's most useful tools, called the periodic table. The periodic table shows all the elements in rows, called periods, arranged in ascending order of their atomic number – that is, the number of protons in their nucleus.

What is remarkable about the table is that all the elements in the vertical columns, called groups, have very similar properties. Sodium and potassium, for instance, both fizz and burn in contact with water and have many properties in common with other elements in Group 1. Group 8, on the other hand, includes all the non-reactive (inert) gases. So knowing how some elements in a group behave allows chemists to predict how other elements in the same group will behave.

The system works essentially because it ties in with the electron shell structure of the atoms. The periods correspond, to some extent, with the various shells – there are seven shells and seven periods. Similarly, the groups correspond to the number of electrons in the outer shell – there are up to eight electrons and eight groups.

The complication comes with the transition elements, a large block in the middle of the table. On the left are the two groups of metals, with one and two electrons in their outer shells. To the right are the six groups of poor metals, metalloids, and non-metals. The transition elements do not fall into any group, but between groups 2 and 3. This is because they never have more than two electrons in their outer shell. Electrons fill out the next shell in from the outer shell instead.

Sodium, a metal, forms common salt with chlorine. Sodium lamps are used for street lighting.

Na¹¹ Sodium 23

Calcium, an alkaline earth metal, is vital to forming strong teeth and bones. One of the best dietary sources of calcium is a dairy product such as milk.

Ca Calcium 40

Chemical symbol

Cu²⁹

Copper
64

Atomic number

Name of element

Relative atomic mass

Reactive metals

	Group 1	Group 2
Period 2	Li ³ Lithium 7	Be ⁴ Beryllium 9
Period 3	Na ¹¹ Sodium 23	Mg ¹² Magnesium 24
Period 4	K ¹⁹ Potassium 39	Ca ²⁰ Calcium 40
Period 5	Rb ³⁷ Rubidium 85	Sr ³⁸ Strontium 88
Period 6	Cs ⁵⁵ Caesium 133	Ba ⁵⁶ Barium 137
Period 7	Fr ⁸⁷ Francium 223	Ra ⁸⁸ Radium 226

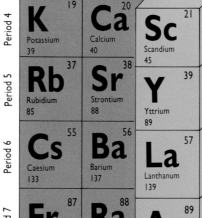

Sc ²¹ Scandium 45	
Y ³⁹ Yttrium 89	
La ⁵⁷ Lanthanum 139	
Ac ⁸⁹ Actinium 227	

Transition metals

Ti ²² Titanium 48	V ²³ Vanadium 51	Cr ²⁴ Chromium 52	Mn ²⁵ Manganese 55
Zr ⁴⁰ Zirconium 91	Nb ⁴¹ Niobium 93	Mo ⁴² Molybdenum 96	Tc ⁴³ Technetium 98
Hf ⁷² Hafnium 178	Ta ⁷³ Tantalum 181	W ⁷⁴ Tungsten 181	Re ⁷⁵ Rhenium 186
Ku ¹⁰⁴ Kurchatovium 261	Ha ¹⁰⁵ Hahnium 262	¹⁰⁶ 263	

The reactive metals include the alkali metals of Group 1, such as lithium, sodium, and potassium, and the alkaline earth metals of Group 2, such as beryllium, magnesium, and calcium. They are all silvery-colored and tarnish rapidly in air. They also tend to be fairly reactive, especially toward the bottom of Group 1. Group 1 metals react with water to form alkaline solutions such as potassium hydroxide or caustic soda, used in soap making.

The transition metals are generally hard, tough and shiny – unlike the softer reactive metals. They conduct electricity and heat well, and have high melting points. These properties make them durable and malleable,

Inner transition metals

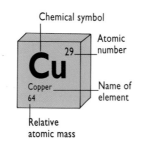

	Ce ⁵⁸ Cerium 140	Pr ⁵⁹ Praseodymium 141	Nd ⁶⁰ Neodymium 144	Pm ⁶¹ Promethium 147	Sm ⁶² Samarium 150	Eu ⁶³ Europium 152	Gd ⁶⁴ Gadolinium 157	Tb ⁶⁵ Terbium 159
Lanthanides								
Actinides	Th ⁹⁰ Thorium 232	Pa ⁹¹ Protactinium 231	U ⁹² Uranium 238	Np ⁹³ Neptunium 237	Pu ⁹⁴ Plutonium 242	Am ⁹⁵ Americium 243	Cm ⁹⁶ Curium 247	Bk ⁹⁷ Berkelium 247

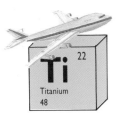

Titanium is often mixed with other metals to make strong, light alloys, commonly used in aircraft components.

Ti	22
Titanium	
48	

Sulphur is a non-metal that is often used as a fungicide on plants.

S	
Sulphur	
32	

H	1
Hydrogen	
1	

Silicon, one of the most abundant elements, is used to make the chips that control every computer.

Si	14
Silicon	
28	

Non-metals

Group 8

He	2
Helium	
4	

Period 1

Group 3 Group 4 Group 5 Group 6 Group 7

B	5	C	6	N	7	O	8	F	9	Ne	10
Boron		Carbon		Nitrogen		Oxygen		Fluorine		Neon	
11		12		14		16		19		20	

Period 2

Al	13	Si	14	P	15	S	16	Cl	17	Ar	18
Aluminium		Silicon		Phosphorus		Sulphur		Chlorine		Argon	
27		28		31		32		35		40	

Period 3

Fe	26	Co	27	Ni	28	Cu	29	Zn	30	Ga	31	Ge	32	As	33	Se	34	Br	35	Kr	36
Iron		Cobalt		Nickel		Copper		Zinc		Gallium		Germanium		Arsenic		Selenium		Bromine		Krypton	
56		59		59		64		65		70		73		75		79		80		84	

Period 4

Ru	44	Rh	45	Pd	46	Ag	47	Cd	48	In	49	Sn	50	Sb	51	Te	52	I	53	Xe	54
Ruthenium		Rhodium		Palladium		Silver		Cadmium		Indium		Tin		Antimony		Tellurium		Iodine		Xenon	
101		103		106		108		112		115		119		122		128		127		131	

Period 5

Os	76	Ir	77	Pt	78	Au	79	Hg	80	Tl	81	Pb	82	Bi	83	Po	84	At	85	Rn	86
Osmium		Iridium		Platinum		Gold		Mercury		Thallium		Lead		Bismuth		Polonium		Astatine		Radon	
190		192		195		197		201		204		207		209		210		210		222	

Period 6

Poor metals

Metalloids

Un	109
Unnilennium	

which is why they are useful for making things. Valuable metals such as titanium, iron, silver, gold, and copper are all transition metals. Titanium, for instance, is used to make light, strong corrosion-resistant alloys for everything from aircraft wings to heart pacemakers.

Lanthanides *are sometimes called the rare earths. They are not often used in everyday life. They are silvery reactive metals such as ytterbium.*

Actinides *are radioactive metals, most of which are created artificially. Only the first four in the series, including uranium, occur naturally.*

The poor metals *are similar to the true metals, but often do not have all the characteristics of the true metals. The poor metal aluminum is widely used – especially when mixed in an alloy – as a component in structures where lightness is at a premium.*

Metalloids *have things in common with metals and non-metals. They conduct electricity only in certain conditions, and this makes them valuable to the electronics industry where some are used in transistors, and semiconductors.*

The non-metals *are usually gases, such as helium and argon. Even when solid, like sulfur, they are dull and often brittle, not shiny and hard like metals. Non-metals are poor conductors of electricity and melt at low temperatures compared with metals. They also form compounds with oxygen (oxides) which are acidic or neutral, not alkaline like the metals.*

Uranium is used as the fuel in nuclear power stations.

U	92
Uranium	
238	

Dy	66	Ho	67	Er	68	Tm	69	Yb	70	Lu	71
Dysprosium		Holmium		Erbium		Thulium		Ytterbium		Lutetium	
163		165		167		169		173		175	

Cf	98	Es	99	Fm	100	Md	101	No	102	Lr	103
Californium		Einsteinium		Fermium		Mendelevium		Nobelium		Lawrencium	
249		254		253		256		254		257	

From atoms to molecules

Mix chemical elements together and they often form new molecules. What makes them react – and what happens when they do?

When different substances meet, they may have no effect on each other, or they may intermingle to form a mixture or a solution. But the substances may react together in such a way that some of the bonds between atoms are broken and new ones are made, making new compounds. It is this breaking and making of bonds that defines a chemical reaction.

Chemical reactions, some natural, some arranged by humans, are going on in the world around us all the time. Candles burning, nails getting rusty, cakes rising in the oven, and food being digested – all these are chemical reactions. Some involve just two substances; some involve many more. But whenever a reaction takes place, at least one of the substances involved is changed, often irreversibly.

In a chemical reaction, the atoms themselves do not change – the element carbon always remains carbon, and oxygen always remains oxygen – though electrons can be knocked off, added on, or shared between atoms. However, some or all the elements involved may make new molecules. When hydrogen and oxygen gas mix, for instance, a spark makes them react together to form water, which is a compound of oxygen and hydrogen; a molecule of water contains two hydrogen atoms and one oxygen atom. The formation of this new compound – this different molecule – shows a reaction has occurred.

Indeed, one way to describe a chemical reaction is to list the atoms and molecules before the reaction, and the atoms and molecules afterward. Since atoms can normally be neither created nor destroyed, there are always the same atoms after the reaction as before – but they can still make new molecules. Among the most common chemical reactions are oxidation reactions, like the

A fireworks display depends on atoms combining together to make compounds. In the types of fireworks seen at public displays, the bright white flames are likely to be the result of the combustion (burning) of the metal magnesium, in which a great deal of energy is released. During combustion, magnesium combines with oxygen to form magnesium oxide, and the sky is lit up by the light energy given off by the reaction. Different colored flames are due to other metals burning.

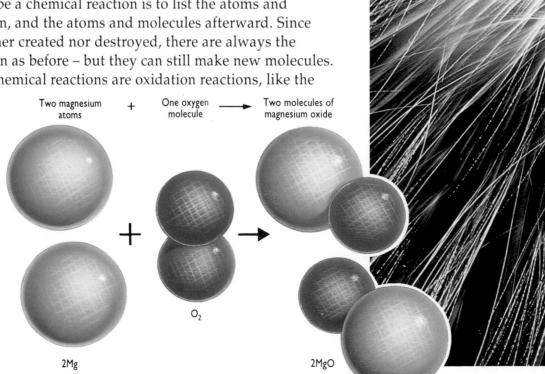

Two magnesium atoms   +   One oxygen molecule ⟶ Two molecules of magnesium oxide

2Mg   + O_2 →   2MgO

MAKING MOLECULES

Some molecules, such as helium molecules, are made up of just a single atom, but most are made from two or more atoms. Molecules of hydrogen and nitrogen are made from their own atoms. But identical atoms may be combined so that they are effectively different substances. This is called allotropy. Ordinary oxygen and ozone (**below**) are allotropes of the element oxygen.

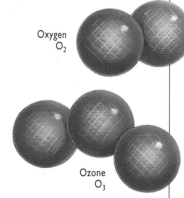

Oxygen
O_2

Ozone
O_3

rusting of iron and the tarnishing of copper. Oxidation once just meant any reaction in which a substance gained oxygen atoms; now it also means any reaction in which a substance loses hydrogen or even electrons. The opposite of oxidation is reduction. Both oxidation and reduction reactions are essential to most life processes.

A cake rises in the oven because chemical reactions in the cake batter create tiny gas bubbles that inflate it like a host of miniature balloons. In some cakes, the gas is made by yeast, a brew of tiny living cells that turns sugar, of one sort or another, into water and the gas carbon dioxide, which gives rise to the bubbles. In other cakes the gas is made by baking soda, added in small quantities to the cake batter. (Self-rising flour is simply flour with the baking soda already added.) Baking soda is a mixture of tartaric acid powder and bicarbonate of sodium (called sodium hydrogencarbonate). When the cake is put in the oven, the acid dissolves in the moisture of the cake, and the heat makes the bicarbonate react with it to produce bubbles of carbon dioxide gas and water.

Most molecules of elements are, like hydrogen, made of the same atoms. But molecules of compounds are made from atoms of two or more elements – always in the same combination. A methane molecule is made from one carbon and four hydrogen atoms. Water is made of one oxygen and two hydrogen atoms, and carbon dioxide is made of one carbon and two oxygen atoms.

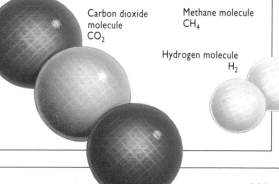

Water molecule
H_2O

Carbon dioxide
molecule
CO_2

Methane molecule
CH_4

Hydrogen molecule
H_2

Key to atoms

 Oxygen   Carbon   Hydrogen

The intimate bonds

Molecules are made when atoms become bonded together.
But just what is it that makes these bonds?

Atoms seem unlikely partners. Their electron clouds make them negatively charged on the outside. Since like charges repel, atoms might be expected to repel each other. The key to bonding lies in the atom's electron shell structure.

One way bonding can happen is by the sharing of electrons between atoms with one or more electrons "missing" from their outer shells. This process is called covalent bonding and happens between the atoms of elements called the non-metals and transition metals in the periodic table.

When two such atoms meet, the attraction of the nucleus of each atom is strong enough to draw an electron (or electrons) from the other atom into its orbit to fill it up. Hydrogen atoms each have a single electron so when two hydrogen atoms meet, their electron clouds intermingle and they share electrons, as if each had the full two. This is why hydrogen atoms usually go around in pairs.

Covalent compounds are formed when two or more different atoms share electrons in this way. For instance, oxygen has two gaps in its outer shell, and in the compound water an oxygen atom shares two of its six electrons with two hydrogen atoms. In return it borrows one from each hydrogen atom, so all have full outer shells.

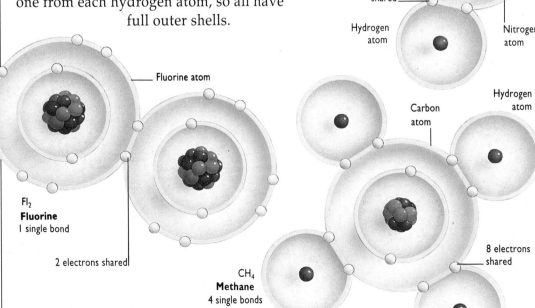

Fl_2
Fluorine
I single bond

Fluorine atom

2 electrons shared

CH_4
Methane
4 single bonds

Carbon atom

Hydrogen atom

Hydrogen atom

8 electrons shared

NH_3
Ammonia
3 single bonds

6 electrons shared

Hydrogen atom

Nitrogen atom

Plants need nitrogen to grow, but few can extract it from air because covalently bonded nitrogen molecules are stable and unreactive. Instead, plants get nitrogen from nitrogen compounds in the soil. The most important is ammonia (above left), a covalent molecule made of one nitrogen and three hydrogen atoms. Ammonia fertilizers dramatically boost crop yields – especially wheat (above).

All chlorine molecules share one electron, making a single bond (far left). Oxygen molecules (right) share two, making a double bond. Carbon, lacking four electrons, forms two double bonds with oxygen to make carbon dioxide (far right). Like many covalent compounds, carbon-based methane (left) occurs naturally as a gas because covalent bonds are so strong that the molecules are self-contained and unwilling to form the links that hold solids and liquids together.

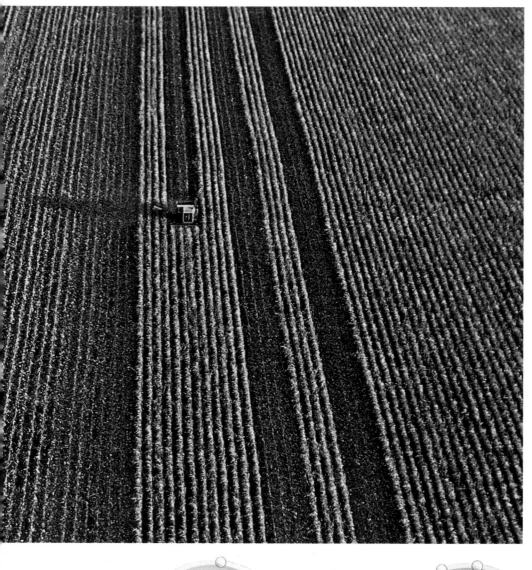

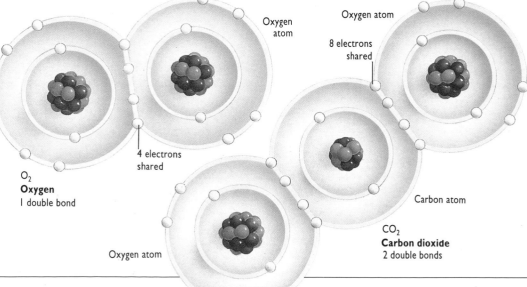

O₂
Oxygen
I double bond

4 electrons shared

Oxygen atom

Oxygen atom

8 electrons shared

CO₂
Carbon dioxide
2 double bonds

Carbon atom

Oxygen atom

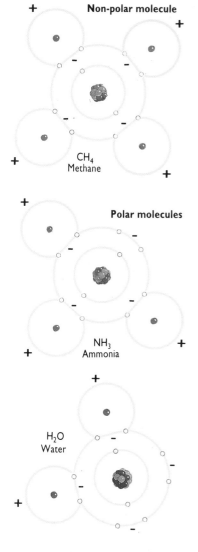

+ **Non-polar molecule** +

− −

CH₄
Methane

+ +

+ **Polar molecules**

−

−

NH₃
Ammonia

+ +

+

H₂O
Water

−

+

POLAR MOLECULES

If, when atoms share electrons, the nucleus of one atom is larger than the other, it draws electrons nearer to it. This is called a polar bond. In the carbon-hydrogen bonds in methane (**top**), the electrons are drawn nearer to carbon than hydrogen. In polar molecules, such as ammonia and water (**above**), not just single bonds are polarized but the whole molecule, because one end is more negatively charged than the other.

Metal connections

In the giant crystal structures that metal atoms form, electrons are free to wander. This helps explain how metals conduct current.

There are over 80 kinds of pure metal – well over four-fifths of the total of the basic chemical elements of the periodic table – and their unique combination of qualities makes them among the most useful of all substances. They are strong, hard, and easy to shape into anything from a table fork to an aircraft. They have high melting and boiling points, so metal structures can withstand high temperatures. And they conduct electricity and heat better than any other solid. Copper, especially, is an extremely good conductor of electricity, which is why it is used for making most electrical wires and cables.

This combination of qualities in the metals comes from the way their atoms are bonded together. In bonds between metal atoms, electrons are shared between the atoms which makes them rather like the covalent bonds between molecules. But instead of forming molecules, metals form giant crystal structures. If they are dipped in acid or looked at under a microscope, this structure can sometimes be seen.

Metal atoms all have just a few electrons in their outer shells, and these are only loosely held in place. When two metal atoms are close together, their outer shells overlap, and the loose electrons

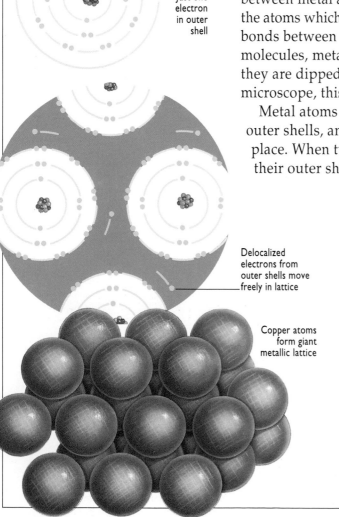

Copper atom

Just one electron in outer shell

Delocalized electrons from outer shells move freely in lattice

Copper atoms form giant metallic lattice

*The exterior of the Statue of Liberty (**right**) is a cladding made of copper metal. This cladding is supported by stainless steel rods which give the structure strength. The monument, which has welcomed generations of immigrants into the United States, survived a major rebuild in the 1980s.*

*A typical metal, copper (**left, above**) has only one electron in its outer shell, which it readily donates to the free-ranging cloud of electrons formed when metal atoms get together (**left, center**). The electrons from many metal atoms orbit freely around between the atoms.*

*The strength of metallic bonds lies in the powerful attraction between the negatively charged free electrons and the metal cations – the positively charged ions left behind when the atoms lose these electrons. These bind the particles together in a giant metallic lattice (**left, below**).*

tend to orbit around both nuclei. If a third atom approaches, it too gets drawn into the association, and its loose electrons join the others circling freely around the other nuclei. In this way, huge, closely packed structures of atoms build up with loose or "free" electrons wandering around between the atoms in a shared cloud. Atoms that have lost or gained electrons are called ions, so the proper name for the metal atoms in this association is, in fact, ions, since all of these atoms have lost an electron or two to the free-floating cloud that swathes the crystal lattice formed when metal atoms accumulate.

This remarkable structure means metals are strong and can often be bent without shattering because the atoms slide over each other and re-form in their new positions. The free electrons make metals good electrical conductors – a difference in electrical charge between the two ends of a piece of metal encourages the free electrons to move away from the negative charge, and toward the positive charge. The free electrons also help to conduct heat energy quickly.

Metals, like all elements, have set melting points and one – mercury – is actually liquid at room temperature. Usually, however, the bonds in solid metals are hard to break. This is demonstrated by the extraordinarily high temperature of 6,170°F (3,410°C) needed to melt the metal tungsten, which is used in glowing light bulb filaments.

THE SHINING

The shininess of metals has always been one of their main attractions – especially of gold and silver (**above**). The shininess stems from the huge numbers of free electrons ready and waiting to be excited. When light falls on metals, it temporarily stimulates a wide range of the electrons which then re-emit light, making the metal shine brightly.

See also

ATOMS AND MATTER
▶ Inside the atom
104/105

▶ The key to behavior
106/107

▶ Classifying the elements
108/109

▶ Attraction of opposites
116/117

▶ Manipulating metals
132/133

▶ Light from atoms
138/139

ENERGY
▶ Generating electricity
68/69

▶ A matter of degree
80/81

ELECTRONS: THE PARTICLES OF POWER

The key to a metal's electrical conductivity lies in the free electrons. If there is an electric field between one end of a piece of metal and the other – that is, if one end is more positively charged than the other – free electrons drift away from the negatively charged end. The individual electrons hardly move at all, but the charge is transmitted through the metal by a knock-on effect from electron to electron.

The ability of metals to carry an electric current is used in hundreds of different ways. For instance, copper overhead cables carry the huge amounts of electrical power needed to drive an electric locomotive over considerable distances. But the current will not flow unless the circuit is unbroken, which is why the locomotive has spring-loaded cable followers, called pantographs, to keep it in continuous contact with the overhead supply.

No current flowing

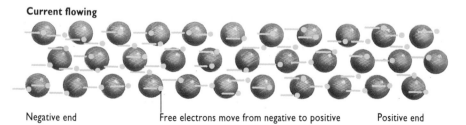

Copper atom   Free electrons move randomly

Current flowing

Negative end   Free electrons move from negative to positive   Positive end

Attraction of opposites

Despite their different characteristics, atoms of metals and non-metals can be extremely attractive to each other.

Metals and non-metals can bond together by transferring electrons in what is called an ionic bond. Anything from individual molecules to giant crystal networks can be linked up this way when atoms exchange electrons.

When the metal sodium and non-metal chlorine atoms meet, for instance, they may combine to form sodium chloride, or common salt. A chlorine atom has seven electrons in its outer shell, one short of the stable eight of inert gases. A sodium atom has an outer shell with just one electron in it above a "full" shell of eight. As they meet, the sodium's "spare" electron migrates to the chlorine, giving both atoms a full eight in their outer shells. This makes the chlorine negatively charged, since it now has an extra negatively charged electron, and the sodium positively charged, since it has "lost" one electron.

The charge on the atoms, or ions, is the key to ionic bonds. Opposite charges attract, so negative chlorine ions (anions) and positive sodium ions (cations) are drawn to each other and bond together, arranging themselves into salt crystals.

Ionic or electrovalent bonds can form like this whenever chemicals react together to make ions. Metal atoms with "spare" electrons can lose them to form cations; non-metal atoms lacking electrons can gain them to form anions. When magnesium in fireworks burns, it gives the two electrons in its outer shell to oxygen, which has six electrons in its outer shell, to form magnesium oxide.

Ionic crystals such as salt are fairly strong, but dissolve easily in water. When this happens, the bonds between the ions weaken, and they separate to drift

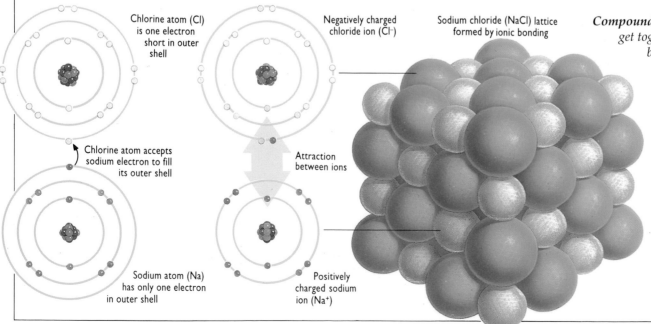

Chlorine atom (Cl) is one electron short in outer shell

Negatively charged chloride ion (Cl⁻)

Sodium chloride (NaCl) lattice formed by ionic bonding

Chlorine atom accepts sodium electron to fill its outer shell

Attraction between ions

Sodium atom (Na) has only one electron in outer shell

Positively charged sodium ion (Na⁺)

*Compounds formed when metals and non-metals get together rely on electron exchange. The type of bond formed is called an ionic bond because, after the electron exchange, the atoms of the elements involved become ions – charged particles. In common salt (**left**), sodium, a metal with one electron in its outer shell, donates the electron to the non-metal chlorine, which has one short of the full complement of eight electrons in its outer shell. Solid salt is a crystal, and the ions stack up in a regular pattern or lattice.*

Rock salt, found underground, is the source of most of our common salt. The most familiar use of salt is to flavor food, but it is also used in soap manufacture, for keeping roads free of ice in winter, and as a source of chlorine.

*The columns of salt in the Dead Sea, in the Middle East (**left**), are made mostly of the ionically bonded compound sodium chloride, or common salt. It is one of the most widespread and most useful of all ionic compounds.*

*Silver-plated dishes actually contain very little silver (**below**). They are usually made from cheaper metals such as nickel, copper, and zinc, with a thin coating of silver. The silver coating is put on by electrolysis, which is why silver-plated articles are sometimes marked EPNS, standing for electroplated nickel silver.*

The item to be plated is dipped into a solution of a silver salt and made the cathode (negative terminal) of an electric circuit. When the current is switched on, positively charged silver ions migrate to the cathode, pick up electrons, come out of the solution as metal, and stick to the item. The result is a beautifully even but very thin coat of silver.

freely through the water. Because these ions are mobile and electrically charged, the solution conducts electricity well.

For instance, if electric terminals are dipped in a salt solution, positively charged sodium cations are drawn to the negative terminal (cathode) where the current creates excess electrons; chlorine anions are drawn to the positive terminal (anode), where there is a shortage of electrons. The sodium cations take up excess electrons from the cathode to become sodium metal atoms; the chlorine ions lose their extra electron to the anode to become chlorine gas atoms. This electrical splitting of an ionic compound is called electrolysis. Batteries reverse the process to create an electric current by using terminals made from two different metals. The result is that atoms from the cathode are drawn as cations into solution. An excess of electrons is left behind, and an electric charge created.

117

Heated reactions

Every type of chemical reaction, from boiling an egg to burning wood, involves a change in heat energy.

Just as a certain amount of effort is needed to pry apart the halves of a walnut, so energy is needed to break bonds between atoms – and making new bonds releases this energy. This is why every chemical reaction involves a change in energy. Some reactions, such as those that go on in a battery, involve electrical energy; others involve light or sound, but most chemical reactions involve heat energy.

Much of the energy in chemical reactions comes from the internal energy of the substances themselves. There is energy locked up in the chemical bonds, for instance, and in the orbiting electrons, and there is energy in the vibration and movement of the particles. The sum total of all this internal energy is called enthalpy.

In some reactions, making new bonds releases more energy than it takes to break the old ones, so the enthalpy of the substances is reduced and the end-products have less internal energy altogether than the reacting substances. If this happens, the excess energy is usually given off as heat. Reactions like this are said to be exothermic. The warmth of a log fire, the heat of a gas flame, and the heat when certain glues mix are all exothermic reactions. For instance, when hydrogen and oxygen combine to make water, the reaction gives out a great deal of heat to the surroundings.

The energy we need to move *comes from exothermic chemical reactions in the body. At the heart of the process are molecules of the sugar glucose, the body's fuel. Glucose is "burned" in the muscles to release its energy, in the same way as any other fuel is burned – except the burning takes place at low temperatures. When the glucose burns, it combines with oxygen in an exothermic reaction called* respiration. It is the heat of this reaction that makes a runner hot.

The body gets most of its glucose by processing carbohydrates in food in the liver. The liver either releases glucose into the bloodstream or stores it until it is needed. In muscles, the carbon in glucose combines with oxygen, carried in blood from the lungs, to form carbon dioxide, which is carried back to the lungs and expelled.

The burning of rocket fuel is a violent exothermic reaction, producing great quantities of energy, heat, light, and sound. Like most forms of combustion, it relies on elements within the fuel combining with oxygen as they burn. But since there is no oxygen in space, rockets must carry their own with them, usually in liquid form. Huge amounts of fuel and oxygen are needed for a rocket launch: a rocket such as the American *Saturn V* burns 15 tons of kerosene/oxygen mix every single second for the first two and a half minutes of its flight.

If there is not enough internal energy to break all the bonds, heat may be drawn in from the surroundings. This is called an endothermic reaction. When carbon and hydrogen are mixed to make acetylene, a gas that is used in welding, much heat energy is drawn in from the surroundings as the gas is made.

Endothermic reactions happen when food is cooked. As you cook, bonds in the food are made and broken. The net result is a gain in the internal energy of the ingredients so energy (heat) is needed to make the reaction happen. When an egg is fried, the change in bonds is evident as the egg solidifies.

Mints taste cool because they create an endothermic reaction – a reaction which draws in heat from its surroundings – when they encounter the moisture in your mouth. The saliva produced as you suck a mint triggers the chemical reaction, and for it to work it has to take in heat from some source. Heat from your tongue, palate, and mouth enables the reaction to proceed. The energy thus supplied is used in the making and breaking of chemical bonds in the reaction.

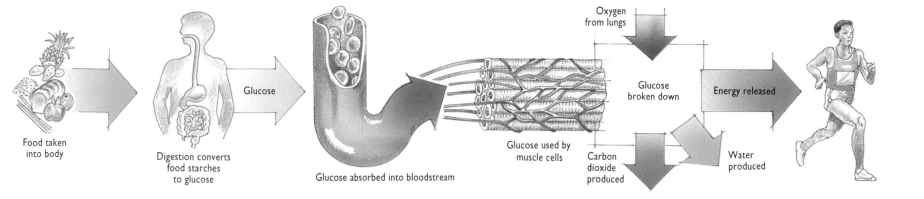

Food taken into body

Digestion converts food starches to glucose

Glucose absorbed into bloodstream

Glucose

Glucose used by muscle cells

Oxygen from lungs

Glucose broken down

Carbon dioxide produced

Energy released

Water produced

Atomic shorthand

Words, letters, numbers, symbols, and models are used to describe chemicals and their three-dimensional structure simply and clearly.

The best known way of describing a chemical is with its chemical formula. This uses letters and numbers to show all the atoms that make up a substance. Letters represent different elements, and numbers after a letter show how many atoms of this element are in a molecule of a substance.

Thus the chemical formula for a molecule of hydrogen is H_2, because it consists of two atoms of hydrogen; and the formula for water is H_2O, because water molecules are made from two atoms of hydrogen and one atom of oxygen.

One benefit of formulae is that they can describe chemical reactions neatly in equations. The reaction of sodium (Na) and water – to produce sodium hydroxide (NaOH) plus hydrogen – looks like this: $Na + H_2O \rightarrow NaOH + H_2$. This shows just the substances involved, not how many atoms and molecules take part. So balanced equations are used, in which a number before each formula indicates how many molecules are involved, like this: $2Na + 2H_2O \rightarrow 2NaOH + H_2$.

MIRROR MOLECULES

Molecules of a compound are usually identical. But a few organic chemicals come in two versions: right-handed and left-handed. They have the same arrangement of atoms, but are put together the opposite way round like left and right hands; they are called enantiomers. These mirror-image molecules, such as the two types of lactic acid in milk, usually have the same properties, but can react at different rates and in different ways with other enantiomers.

A chemical's name and formula identify it, but they do not tell us about its structure. To describe how it looks, scientists use written shorthand, structural models, and ball and stick models.

Written shorthand and the more elaborate structural model show the structure of molecules simply by placing the letter for each atom of the compound in approximately the right place and drawing lines between them to indicate the nature of the bond. Thus a single covalent bond is indicated by a single line, a double bond by a double line, and so on. In written shorthand, shortcuts are taken, and the structure of certain compounds or parts of compounds is shown just as a series of lines, as in the six carbon atoms that make up the benzene ring.

Ball and stick models are a simple way of showing the three-dimensional structure of compounds. Balls represent atoms and sticks the bonds between them. Thus the ball and stick model (right) of methane, which is carried from its production sites in pipelines (far right), shows its pyramidal, tetrahedral structure.

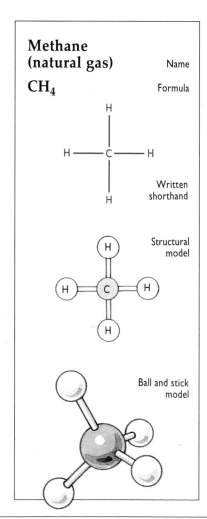

Methane (natural gas) — Name

CH_4 — Formula

Written shorthand

Structural model

Ball and stick model

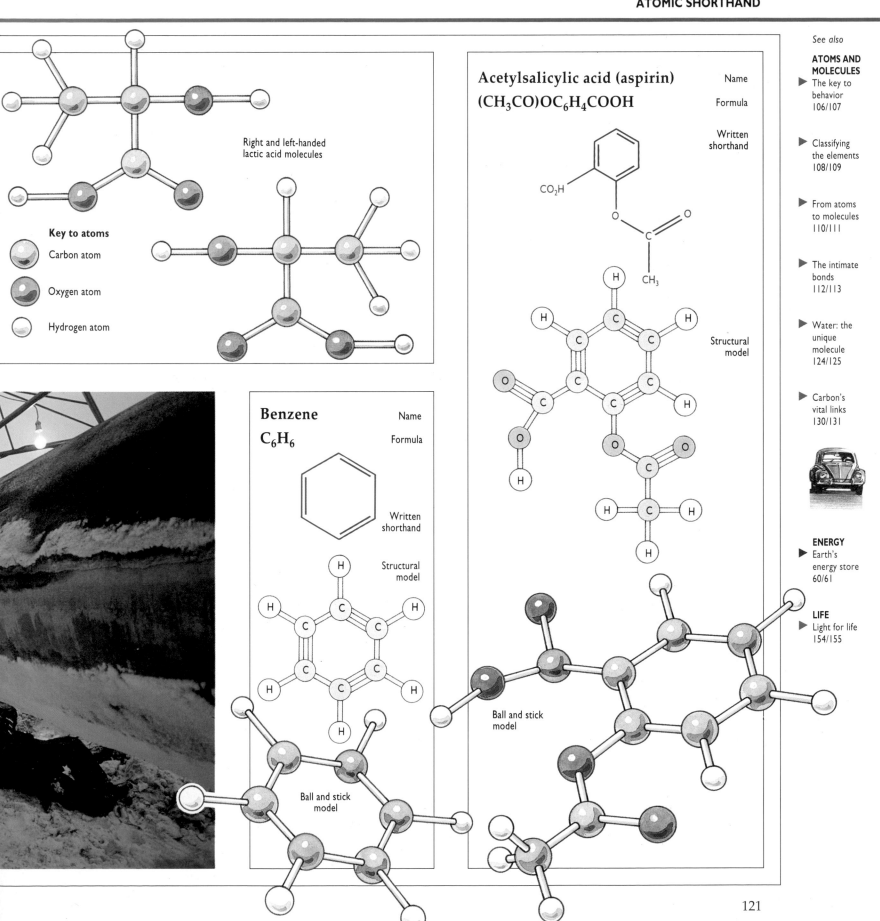

Right and left-handed lactic acid molecules

Key to atoms

Carbon atom

Oxygen atom

Hydrogen atom

Acetylsalicylic acid (aspirin) — Name

$(CH_3CO)OC_6H_4COOH$ — Formula

Written shorthand

CO_2H

CH_3

Structural model

Ball and stick model

Benzene — Name

C_6H_6 — Formula

Written shorthand

Structural model

Ball and stick model

What's the solution?

Mix one substance with another and sometimes they seem to become one and sometimes they refuse to change.

Solutions are among the most important of all mixtures. Every time you make a cup of instant coffee, you are making a solution – a mixture in which the ingredients are so well mixed it seems like a single substance. Solutions can be solid mixed with solid or gas with gas; but most are a solid, liquid, or gas mixed into a liquid.

Every solution has two main elements: the solvent, the liquid that forms the bulk of the solution, and the solute, the substance which dissolves in it. Water is by far the most important solvent, and aqueous (water-based) solutions are the basis of every drink, all the liquids in living organisms, and more besides. But there are also many other non-aqueous solvents, including mineral spirits, nail varnish, gasoline, paint remover, glue, and varnish.

Plain tap water is quite a complex solution. It contains not only a wide variety of dissolved minerals, including both naturally occurring minerals such as calcium salts (which give water its hardness) and those deliberately added for health reasons such as fluoride, but also air. Water that has been boiled tastes flat because all the dissolved air has been boiled out.

Although it seems as though solvent and solute have become one in a solution, they remain chemically separate, and the molecules or ions of the solute simply slot into the spaces between the molecules of the solvent without actually joining up. Solute can be added to the solvent until all the spaces are filled up. Once this happens, no more of the solute will dissolve and the solution is said to be saturated. However, more solute may be dissolved, even in a saturated solution, if the spaces are opened up by heating.

Not all substances are equally soluble in every solvent. Salt, for instance, dissolves in water but not in gasoline. Wax, on the other hand, dissolves in gas

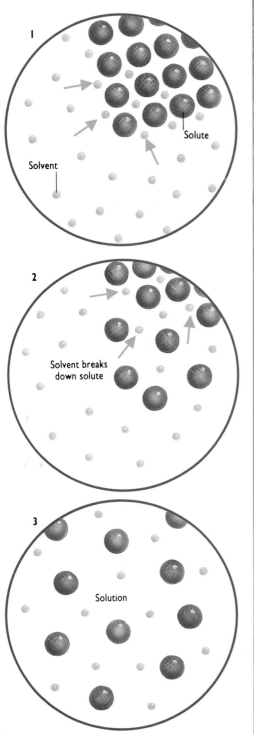

1
Solute
Solvent

2
Solvent breaks down solute

3
Solution

A solution gradually breaks down the bonds that hold a solute together (**1**), and the solute mixes into the solvent (**2**). The two are fully mixed when they are evenly distributed throughout the liquid (**3**).

When a solute is added to or brought into contact with a solvent, the solvent attracts the molecules or ions of the solute out of the structure they are held in as a solid. When common salt dissolves in water, the ions of the sodium and the chlorine that make up the crystal lattice structure of salt are pulled out of the lattice separately by the water molecules. Not only solids dissolve in liquid. Gases can

dissolve in a liquid, as can other liquids like ethanol (alcohol) in water. When one liquid dissolves in another, the liquids are said to be miscible. When this happens, the molecules of the liquids diffuse together. If they do not dissolve, they are said to be immiscible. When two liquids do not dissolve, they may form an emulsion, which is a type of colloid.

One example of an emulsion is the mix of oil and vinegar in a salad dressing. When the dressing is shaken hard, it forms a creamy, thick liquid, but when it stands for a while, the oil and vinegar separate.

Swimming pools are actually filled with quite a complex solution. Water in the swimming pool has a number of extra ingredients to keep it clean and kill germs that might cause infections or diseases. One of the most common substances used for this purpose is the gas chlorine. It deals with dangerous microbes, but also gives water in a pool a distinctive smell and its slight greenish tint.

but not water. When an insoluble substance is added to a liquid, it may mix in other ways. Some particles form a suspension, which means they simply hang in the liquid, like the sand in the waves of sea water at the seaside. Particles so fine they cannot be filtered out form colloids. Milk is a colloid of fat particles held in water. So, too, is detergent foam – it is made from tiny bubbles of gas in a liquid.

Colloids can even form in substances where the molecules of the main substance (called the continuous phase) and the particles mixed in (the dispersed phase) are linked in a lattice. One is gelatin, a colloid of water and animal protein.

SUSPENDED IN THE AIR

Fine particles can be mixed into a gas to form a colloid. Aerosol sprays used to treat sports injuries on the field (**above**) are colloids containing tiny droplets of a pain-killing drug in liquid form (**right**). The active ingredient is pushed out of the spray can through a small nozzle by a propellant gas.

Smoke, too, is a colloid, made from fine soot particles suspended in air. But sand held in the air in a desert storm is a suspension – when the wind drops, the sand falls.

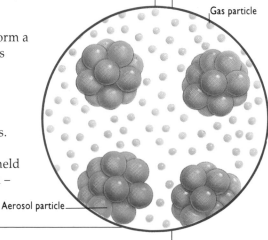

Gas particle

Aerosol particle

Water: the unique molecule

Although it is the most common compound on Earth, water is a most remarkable substance: life would be impossible without it.

Water is everywhere. There are huge quantities of it in the oceans, in the air, in the polar ice caps, on land, and inside every living thing. It plays a part in a vast number of chemical reactions from rusting to photosynthesis.

A remarkable thing about water is that it is commonly found in all three states of matter – solid ice, liquid water, and gaseous water vapor. Similar compounds, such as ammonia and methane, are gases down to -22°F (-30°C) and -256°F (-160°C) respectively, but water does not melt until above 32°F (0°C) and does not boil until 212°F (100°C). Uniquely, water is denser as a liquid than as a solid, which is why it expands when it freezes. When forming ice, water expands by one-eleventh.

*In water molecules (**right**), the positively charged hydrogen ends of the molecules strongly attract the negatively charged oxygen ends of next-door molecules, forming strong bonds, called hydrogen bonds.*

One of the effects this has is to keep water liquid to 212°F (100°C). Because the hydrogen bonds hold water molecules together so tightly, only a few can escape as gas. The bonds also hold water together so well that they create a tense "skin" at the surface.

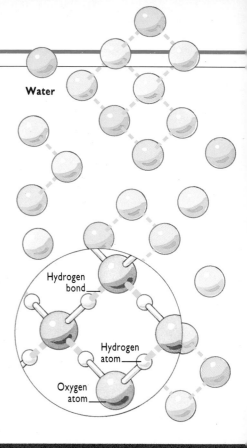

Water

Hydrogen bond

Hydrogen atom

Oxygen atom

*Water has two hydrogen atoms and one oxygen atom (**below**). Each hydrogen shares its electron with the oxygen atom, and shares one of the oxygen's electrons, so it has two electrons to fill its shell. The oxygen shares two of its electrons and two from the hydrogen atoms so it has eight electrons filling its outer shell.*

Oxygen pulls away the electrons from the hydrogen atoms, leaving the positively charged hydrogen nuclei "bare." This means there is a positive charge on the hydrogen end of the V-shaped molecule and a negative charge on the oxygen end. Such a molecule is said to be strongly polar.

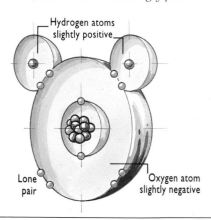

Hydrogen atoms slightly positive

Lone pair

Oxygen atom slightly negative

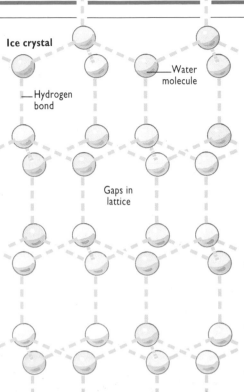

Ice crystal

Water molecule

Hydrogen bond

Gaps in lattice

When water is frozen, the molecules are arranged in a regular but widely spaced lattice (**left**). When it melts, the lattice breaks down, and the hydrogen bonds can pull the molecules together much more tightly. This is why water shrinks, or contracts, when it melts.

Hydrogen bonds are so strong that even when water is heated, they continue to hold the molecules together in clusters of liquid water, creating steam droplets.

Liquid water can also exist below freezing point, a condition known as supercooling. This happens because water freezes only if there is an object, however minute, for it to freeze onto. Once supercooled water is seeded with some impurity, such as dust, its temperature rises to freezing point and it freezes.

WATER OF LIFE

Every body is made up of about three-quarters water. Water-based solutions are vital to every living organism. For instance, each living cell depends on water's ability to dissolve compounds so that essential chemicals can be drawn into and carried out of the cell by a process called osmosis.

Osmosis happens when solutions are different strengths on each side of the cell wall. Water seeps through from the weaker solution to dilute the concentrated solution until both solutions are equal. As it does, it carries chemicals in solution in and out of the cell.

More than 2 percent of the world's water is locked in the frozen ice of the polar regions (**left**). Ice is less dense than water and therefore floats. If ice did not float, the seas would be frozen solid from the bottom up.

Water is at its densest at 39°F (4°C). This means that ice can form across the top of a stretch of water while creatures below can survive in water at a steady 4°C (39°F). As the temperature drops below 39°F (4°C), water expands gradually down to 32°F (0°C) and then expands dramatically as it freezes, with enough force to burst pipes and split rocks.

The acid test

Acids such as lemon juice and alkalis like ammonia make special kinds of solution, each with marked and opposite effects.

Whenever you taste something sour, the chances are you are tasting acid. Oranges, lemons, and other fruit contain citric acid, and vinegar contains ethanoic or acetic acid. Also due to acid is the sting of an ant. Acids such as these are relatively mild; strong acids such as sulfuric acid are so corrosive they can burn through clothes and flesh and even dissolve metals.

Acids are a kind of water solution and it is impossible to tell them apart from any other solution just by looking at them. The only safe way to tell is to test a solution with an indicator, such as red-cabbage water. When an acid is added to red-cabbage water, it turns from purple to bright red. This change can be reversed by adding another kind of solution called an alkali, which cancels out or neutralizes the acid. Adding more alkali eventually turns the solution blue and then green. More refined than cabbage water are universal indicators, chemicals that show a range of colors for a solution's varying acidity and alkalinity.

What makes an acid different from an alkali and every other solution is that it contains many positive hydrogen ions (H^+); that is, hydrogen atoms that have lost their single electrons. In water there are always a few hydrogen ions that have broken away from the water molecules, but only rarely is more than one molecule in 500 million affected. If certain substances are dissolved in water, the number of hydrogen ions increases dramatically to make the solution acidic. When placed in water, hydrogen chloride gas, for instance, splits almost entirely into hydrogen and chlorine ions, making corrosive hydrochloric acid.

In an alkali, there is an excess of negative ions, usually hydroxyl ions (OH^-); that is, water molecules that have lost a hydrogen ion. There are some hydroxyl ions in any water sample, but the numbers become significant only when a substance called a base is added. In water, a base splits into positive, usually metal, ions, and negative, usually hydroxyl, ions. For instance, the base sodium hydroxide, or caustic soda, splits into sodium ions and hydroxyl ions.

Hydrochloric acid

Battery acid

Digestive juices in stomach

Vinegar

Apple juice

Lemon juice

Club soda

Rainwater

Milk

Distilled water

pH 1   2   3   4   5   6   7

← Acidity increases

Neutral

Rain water is always slightly
acidic because carbon dioxide in
the air dissolves in it to form weak
carbonic acid – the acid in club
soda. However, high emissions of
sulfur dioxide from burning fossil
fuels and nitrogen oxides from car
exhausts have increased air pollution
rapidly. Around the world's
industrialized areas, this pollution
is making the rain up to 500 times
as acidic as "normal" rainfall. This
"acid rain" is clearly damaging
trees and killing freshwater life.
Lakes turned acid by acid rain can
be neutralized for a while by adding
lime to them, but the only real
solution is to reduce the pollution.

Acidity and alkalinity are
measured on the pH scale, which is
numbered from 1 to 14 (below). The
midpoint of the scale, called neutral,
is pure, distilled water with a pH of
7. Tap water is slightly alkaline, with
a pH of 8–9; rain water is slightly
acidic with a pH of 5–6, although it
can be higher. As acids become
stronger, the pH number decreases:
stomach acid (pH 2) is ten times
more acidic than an apple (pH 3) and
one hundred times stronger than
vinegar (pH 4). The strongest acids,
such as hydrochloric acid with the
highest proportion of hydrogen ions,
have a pH of 1. At the other end of
the scale the strongest alkalis, such
as potassium hydroxide solution
(caustic potash), which have the most
hydroxyl ions, have a pH of about 14.

See also

**ATOMS AND
MATTER**
▶ Molecules
in motion
98/99

▶ The key to
behavior
106/107

▶ Classifying
the elements
108/109

▶ Attraction of
opposites
116/117

▶ What's the
solution?
122/123

▶ Water:
the unique
molecule
124/125

ENERGY
▶ Generating
electricity
68/69

LIFE
▶ Blueprint
for life
170/171

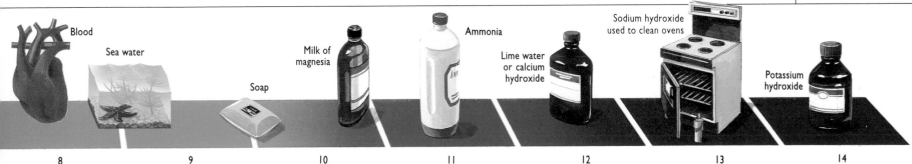

Blood

Sea water

Soap

Milk of
magnesia

Ammonia

Lime water
or calcium
hydroxide

Sodium hydroxide
used to clean ovens

Potassium
hydroxide

8   9   10   11   12   13   14

Alkalinity increases ⟶

Carbon copies

Soot, diamond, charcoal, graphite, carbon fiber – and even the exotically named buckminsterfullerine – are all different, but they have one element in common.

It may seem hard to believe that soot and diamond are chemically identical. Diamond is the world's hardest substance – tough, brilliantly clear, and sparkling, and unaffected by heat until 1,292°F (700°C). Soot is soft, black, and glows at temperatures no hotter than a candle flame. Yet both are made entirely of carbon atoms. The dramatic differences in the varieties of carbon stem from the way their atoms are arranged – in one of two kinds of giant network. One is like that in diamond; the other resembles that in graphite, the carbon compound used in "lead" pencils.

In the diamond network, each atom is linked to four others in an infinitely interlocking tetrahedral pyramidal structure. It is this tight structure that makes diamond so tough. In graphite, however, the atoms are arranged in rings in layers, like sheets of hexagonal tiles. Each atom links with only three atoms, not four as in diamond, and shares a pair of electrons with each – leaving one of the four electrons in its outer shell unpaired. These unattached electrons drift freely between the layers, like the free electrons in a metal, which is why graphite is a conductor of electricity and can look metallic. Charcoal has a graphite-like structure, too, but looks and behaves differently from graphite because it is made up of tiny specks each with just a few sheets of atoms.

A newly discovered form of carbon is buckminsterfullerine, in which atoms link to form stand-alone molecules with between 60 and 960 carbon atoms. These have huge potential as lubricants and superstrong fibers.

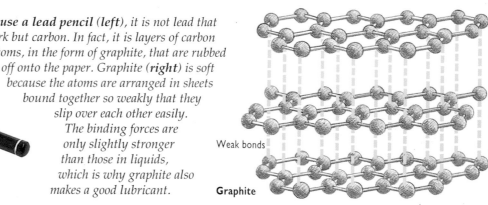

Carbon composites are incredibly strong, light materials, which is why they are perfect for making racing car body shells (right). They are actually a form of graphite which has been spun together to form strong fibers. The main problems are that they are expensive and prone to weaken if their surface is damaged.

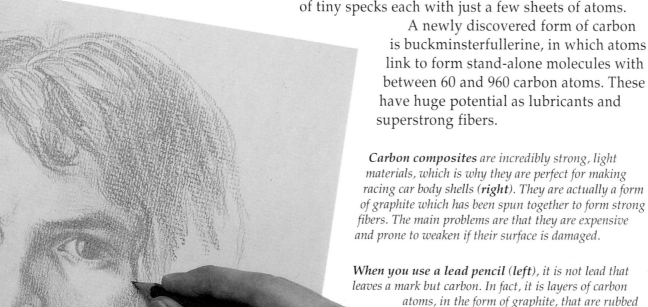

When you use a lead pencil (left), it is not lead that leaves a mark but carbon. In fact, it is layers of carbon atoms, in the form of graphite, that are rubbed off onto the paper. Graphite (right) is soft because the atoms are arranged in sheets bound together so weakly that they slip over each other easily. The binding forces are only slightly stronger than those in liquids, which is why graphite also makes a good lubricant.

Weak bonds

Graphite

C$_{60}$

Diamonds are incredibly hard *because their atoms are arranged in an extremely strong, tightly interlocking pyramidal network (***right***). Graphite can be turned into diamond by squeezing it to push the graphite's layers of atoms closer together until they interlock and make diamond. But the pressure needed to make this happen is enormous, and only small artificial diamonds can be made.*

Diamond

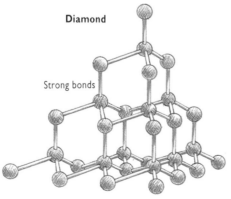

Strong bonds

Buckminsterfullerines, *discovered in 1990, form single molecules, not giant networks. Carbon atoms are linked like graphite, but it is as if the layers have been rolled into shapes such as spirals, tubes, and balls. First to be found was C60 (***right***) – a basketball-shaped molecule of 60 carbon atoms.*

Strong bonds

129

Carbon's vital links

The millions of compounds of carbon include all our fossil fuels, most plastics, and all the materials that make up living things.

Carbon is the most friendly element in the universe. With four electrons in their outer shells, carbon atoms are only too happy to link with other atoms. Compounds of hydrogen and carbon (hydrocarbons) include the fuels oil, coal, and natural gas, and substances such as polyethylene and styrofoam. Many carbon compounds contain other elements. Alcohol contains oxygen, for instance. Chemicals made with carbon are called organic compounds.

Sometimes carbon atoms in organic compounds are linked together in chains (aliphatic compounds); sometimes they are linked in rings (cyclic compounds). Compounds that contain a stable ring of six carbon atoms – called a benzene ring – are called aromatic compounds and include many drugs and dyes. Aromatic compounds are found in a huge number of living plants and animals.

All aliphatic compounds have a backbone of carbon atoms, surrounded by other atoms or groups of atoms. The simplest aliphatic is methane, made from just one carbon atom at the center of four hydrogen atoms, but the carbon backbone can be hundreds of atoms long. New compounds are created as new groups of atoms, called functional groups, such as methyl groups (one carbon and three hydrogen atoms) and amino groups (one nitrogen and two oxygen atoms), slot in place of one or more hydrogen atoms in the basic methane molecule. Chains may build up as the carbon atoms in new methyl groups take the place of hydrogen atoms in methyl groups in the chain. Every addition changes the character of the material.

Oil refineries (above) extract carbon compounds from crude oil.

The behavior of a carbon compound depends on how many carbon atoms link together, the way the carbon atoms link, and, importantly, the type of small group, called a functional group, joined to the chain. Families of molecules with the same functional group are called an homologous series. There are eight main functional groups (left and right).

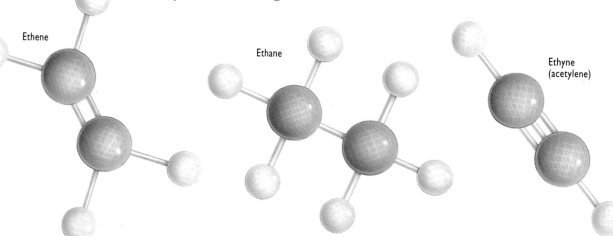

Ethene

Ethane

Ethyne (acetylene)

Alkenes *are hydrocarbons with at least one pair of carbon atoms linked by a double bond. Ethene is used to make vinyl, plastic, and glue.*

Alkanes, *hydrocarbons with most hydrogen atoms and single carbon bonds, range from refinery gas (ethane) to gasoline and bitumen.*

Alkynes, *hydrocarbons with fewest hydrogen atoms and at least some triple carbon bonds, include ethyne, burned with oxygen for welding.*

Key to atoms

Hydrogen

Oxygen

Carbon

Nitrogen

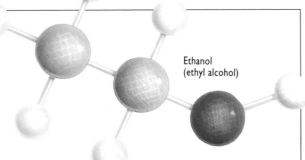

Ethanol
(ethyl alcohol)

Alcohols *contain one or more hydroxyl (hydrogen and oxygen) groups. They include not just ethyl alcohol, the alcohol in intoxicating drinks, but also solvents for gums and resins, oils for perfumes, and many more things.*

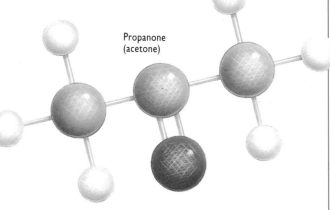

Propanone
(acetone)

Ketones *contain a carbonyl functional group (carbon double-bonded to oxygen) attached to two carbon atoms. The solvent propanone – used to remove nail varnish – is in the ketone family.*

Methanal
(formaldehyde)

Ethylamine

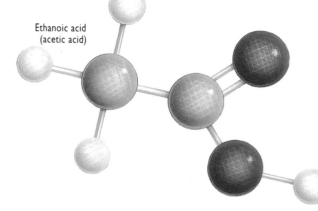

Ethanoic acid
(acetic acid)

Aldehydes *have carbon double-bonded to an oxygen atom. They are basically alcohols with a hydrogen atom removed. They include methanal, used to make the disinfectant formalin.*

Amines *contain an amino functional group (nitrogen and two hydrogen atoms). With carboxylic acids they make amino acids, the building blocks of proteins, the basic chemicals of life.*

Carboxylic acids *contain a carboxyl group (carbon, two oxygens, and a hydrogen) attached to another group. Smaller carboxylic compound molecules include acetic acid (vinegar); larger ones are present in fats in the body.*

131

Manipulating metals

Whether in coins, cutlery, or cars, most metals in everyday use are not pure, but precisely blended mixtures called alloys.

Metals are remarkably useful, but we rarely use them in their pure form. For instance, few people wear pure gold jewelry not because it is expensive, but because it is so soft that it scratches or bends unless handled with care. Usually, gold in jewelry has been hardened by mixing it with small amounts of copper or silver. Most gold jewelry is not 24 carat (100 percent pure) gold, but is 18 carat, a far stronger alloy containing only 75 percent gold.

As with gold, the usefulness of many metals can be improved by blending in small amounts of other elements to make an alloy. In a metal the atoms, or ions, arrange themselves into a regular, crystal-like lattice as the metal solidifies. But crystals made of pure metals can have defects in their lattice structure which make them too soft, or prone to corrode when exposed to moisture and oxygen in the air. In alloying, ions of one metal mix with ions of another metal or metals so the crystal structure is changed, and with it the metal's properties.

Metals in alloys usually need to have similar sized ions and also similar numbers of free electrons so that the ions of one metal can replace ions of the other in the lattice. But while alloys are usually blends of metals, metals can also be alloyed with non-metals. For instance, when a small amount of carbon is added to pure, soft iron, the alloy called steel is made. This extremely strong, versatile material forms because holes in the iron lattice are filled with carbon atoms.

To make most alloys, the elements are mixed when they are molten. Most mixes are then cooled slowly. This allows the ions of the different elements to arrange themselves into a regular lattice structure as the alloy solidifies.

An alloy of aluminum is used in the construction of aircraft because it is light and strong. Aluminum on its own is light, but not strong enough to cope with the severe stresses and huge temperature changes that occur during flight. Today's aircraft are built using an alloy that combines aluminum with small amounts of copper, magnesium, iron, nickel, and silicon. The next generation of planes will probably use an even lighter alloy made from aluminum and lithium.

The world's oldest alloy is bronze, a mixture of copper and tin. Harder than either of its constituents, it revolutionized tool and weapon-making in the Bronze Age, to which it gave its name. Bronze is ideal for casting, because when molten, it flows readily into every part of a mold, faithfully reproducing every detail. Bronze casting is still used today, but some of the most beautiful bronzes, such as this head of the goddess Aphrodite, were created by the ancient Greeks.

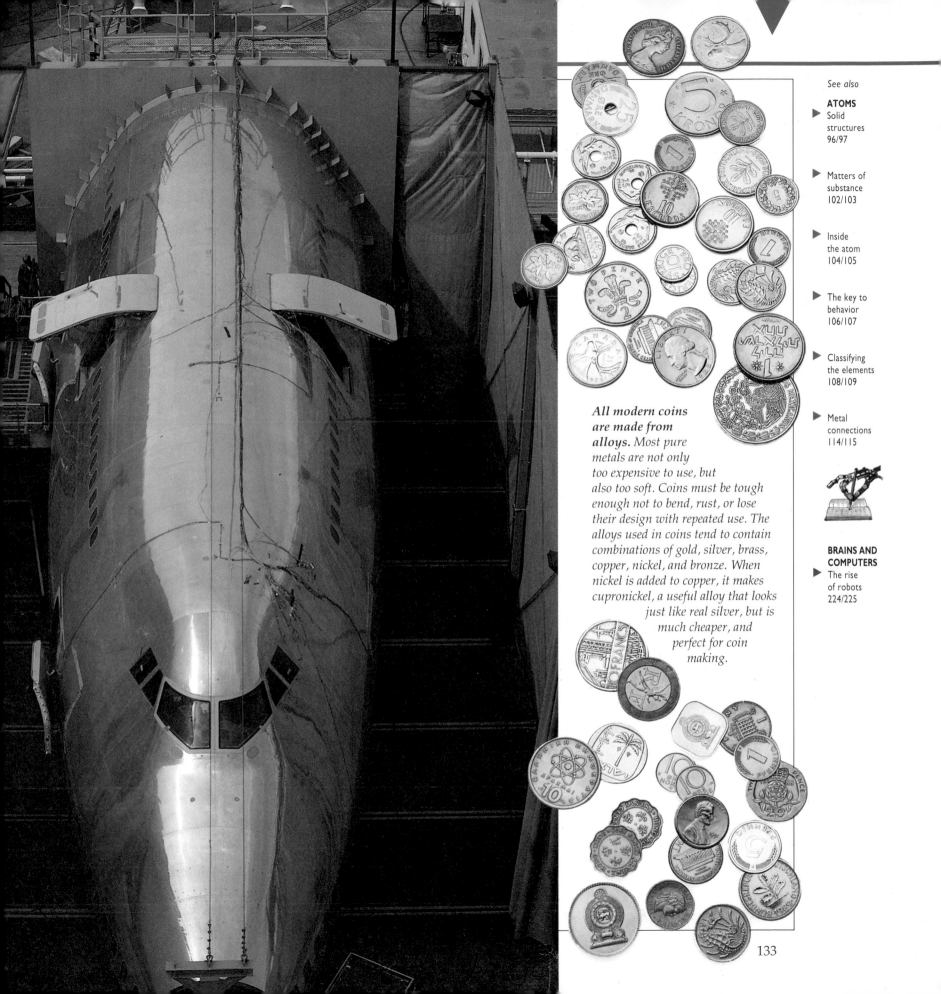

See also

► **ATOMS**
  Solid
  structures
  96/97

► Matters of
  substance
  102/103

► Inside
  the atom
  104/105

► The key to
  behavior
  106/107

► Classifying
  the elements
  108/109

► Metal
  connections
  114/115

**BRAINS AND
COMPUTERS**
► The rise
  of robots
  224/225

*All modern coins
are made from
alloys.* Most pure
metals are not only
too expensive to use, but
also too soft. Coins must be tough
enough not to bend, rust, or lose
their design with repeated use. The
alloys used in coins tend to contain
combinations of gold, silver, brass,
copper, nickel, and bronze. When
nickel is added to copper, it makes
cupronickel, a useful alloy that looks
just like real silver, but is
much cheaper, and
perfect for coin
making.

133

Compounds à la carte

By rearranging molecules, chemists have created a remarkable variety of novel materials – from new drugs to fibers stronger than steel.

Used in everything from kitchenware to spacecraft, plastics are among the most remarkable of all artificial materials – and new uses are being found for them all the time. Almost without exception, they are made from special giant molecules called polymers. Most naturally occurring molecules consist of 20 to 30 atoms at the most, but polymers are made up of hundreds or even thousands of atoms.

The simplest polymers are created when pressure or heat make quite small molecules, called monomers, alter slightly and link together in a long chain – rather like a string of paperclips hooked together. When identical monomers are strung together, addition polymers are made; when two kinds of monomer react together and link up, condensation polymers are made. Polyethylene, for instance, is an addition polymer made from 50,000 or so monomers of a simple hydrocarbon called ethene (also known as ethylene). Some polymers, such as cellulose and silk, occur naturally, but artificial polymers can be tailor-made to suit a wide variety of purposes, and there are now thousands of different kinds.

Stiff-chain polymers such as Kevlar (invented in the 1960s) have been developed which are even stronger than steel and much lighter. Kevlar is now used in everything from skis and other sporting equipment to vital structural components in jet airliners. It is even used for bulletproof vests.

Plastic polymers and plastics were once almost the only

Man-made materials, *constructed by the manipulation of simple chemicals into more complex chemicals, abound. Many of the plastic objects we take for granted today – from shoes to toothbrushes – did not exist just 50 or so years ago.*

There are two types of plastic, the thermoplastics, which get soft when warmed, and the thermoset plastics, which do not soften when warmed.

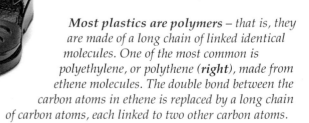

Most plastics are polymers *– that is, they are made of a long chain of linked identical molecules. One of the most common is polyethylene, or polythene (**right**), made from ethene molecules. The double bond between the carbon atoms in ethene is replaced by a long chain of carbon atoms, each linked to two other carbon atoms.*

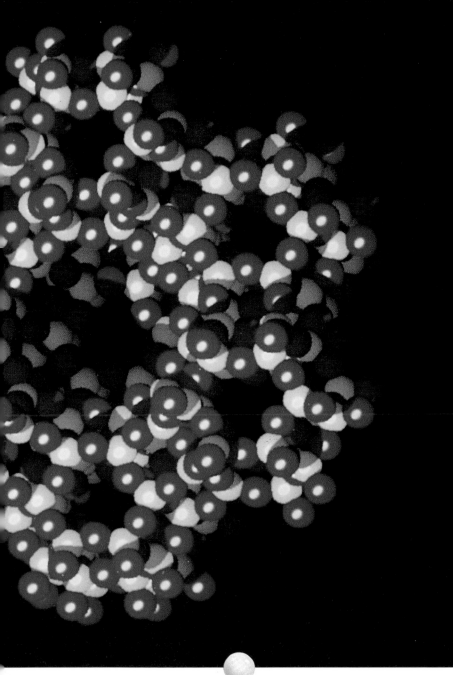

artificial molecules, but since the mid-1960s, chemists have made huge strides in the creation of all kinds of synthetic molecules. Computers have helped enable them to design and test various molecular arrangements quickly on screen. Synthetic molecules now include everything from artificial zeolites used in the petrochemicals industry to liquid crystals used in digital displays.

At the start of the century, chemists dreamed of finding "magic bullets" – specially designed chemicals that would home in on diseased parts of the body and cure them. Now there are hundreds of such drugs, targeted not only at disease-causing organisms, but also at replacing or blocking the effect of certain chemicals within the body to cure a problem.

Drug companies once created new drugs by trial and error, trying out many variations to reach the perfect molecule. Modern computer design means there is no need to actually make many of these compounds, because chemists can put together the molecules and try them for fit on screen.

*Zeolites are naturally occurring minerals made of silicon, aluminum, and oxygen. Remarkably, their molecules fit together to form tiny channels which can trap molecules like a tiny strainer. By changing the ratio of silicon and aluminum, chemists can create synthetic zeolites (**left**) with channels to trap particular molecules.*

See also

ATOMS AND MATTER
▶ Classifying the elements 108/109

▶ From atoms to molecules 110/111

▶ The intimate bonds 112/113

▶ Atomic shorthand 120/121

▶ Carbon's vital links 130/131

LIFE
▶ King carbon 150/151

BRAINS AND COMPUTERS
▶ Modeling reality 226/227

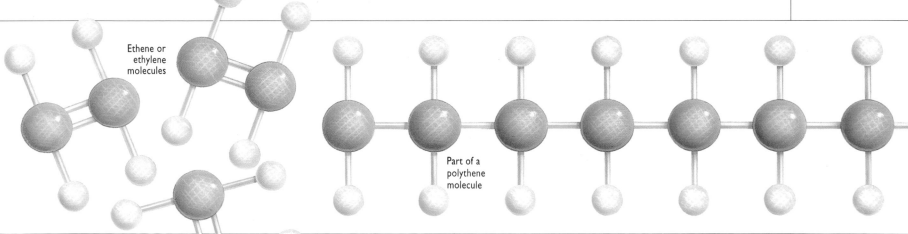

Ethene or ethylene molecules

Part of a polythene molecule

The atomic paintbox

Many things we see in the world around us have a color: they get it by absorbing some wavelengths of light and reflecting others. And some things have no color.

Very few things actually give out light of their own accord; they simply reflect light from another source. So they get their color from the way they interact with the light falling on them, absorbing some wavelengths (colors) of light and reflecting others.

Substances which are colorless – air, white clouds, smoke, and white metals, for instance – cannot absorb any of the wavelengths we can see. Light either bounces straight back off them or passes through. When light strikes colorless solids and liquids, it sets the electrons in the substance vibrating at the same frequencies as the light; the vibrating electrons then re-emit the light virtually unchanged and because of this they appear to have no color.

Colored substances, however, absorb certain colors of light – the rest they reflect. The color they absorb depends on the configuration of the electrons around the atoms. Atoms absorb only light that contains exactly the amount of energy to bump an electron up to a higher energy level – that is, farther from the nucleus.

Surprisingly, visible light rarely has the energy to shift electrons in this way, so only a few substances can actually absorb light and appear colored. Metals such as gold and copper are colored because there are gaps in their penultimate electron shell. Because of this, the energy in visible light can easily move electrons into these gaps. Substances with large, spread-out electron clouds can also absorb visible light. Many natural pigments such as the green chlorophyll in leaves, the red hemoglobin in blood, and the colors in flowers work this way. So, too, do many synthetic dyes.

In most substances, the small quantity of light energy absorbed is dispersed as heat throughout the substance. In some, however, it is re-emitted as light. In glow-paints, this happens instantly; in luminous paint, it is re-emitted a little later. Invisible ultraviolet light is often re-emitted from luminous paint as visible light, which is why the paint appears unnaturally bright. Sometimes, as in lasers, substances will re-emit light of a particular wavelength.

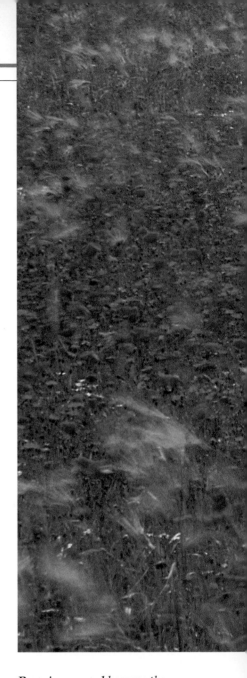

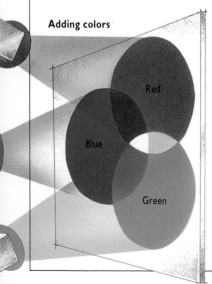

Adding colors

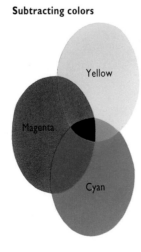

Subtracting colors

*All the rainbow colors we see can be made by mixing just three primary colors of light – red, blue, and green – in different proportions (**far left**). Adding red and blue light, for instance, gives purple light. White light is a mixture of all three colors.*

*Pigments, paints, dyes, and printed colors get their color by absorbing different proportions of red, blue, and green (**left**). So a pure red pigment is one that absorbs blue light and green light; a pure blue pigment is one that absorbs green and red; and a pure green is one that absorbs red and blue light. Combining any two of these would absorb all colors of light, so the primary pigment colors – yellow, cyan (greenish-blue), and magenta (crimson) – are those that absorb just one of the primary colors of light, not two. Yellow pigment absorbs only blue; cyan absorbs only red; magenta absorbs only green.*

*Poppies are red because the flowers (**right**) absorb some wavelengths of the white light falling on them and re-emit others. In fact, the poppy absorbs light from the blue end of the spectrum and re-emits light from the red end of the spectrum.*

A blue flower, such as a bachelor's-button, appears to be blue because it absorbs wavelengths of light from the red end of the visible spectrum and re-emits light from the blue end. Other flowers absorb different wavelengths to give their individual colors.

Only red light emitted

All colors absorbed

White light contains all colors

A simple way to change the color of an object is to paint it. But paint does not actually change the underlying color of an object, it simply coats it with a layer of pigment that absorbs different wavelengths of the light falling on it. Purple paint (**right**), for instance, absorbs most wavelengths from the green and some from the red and blue parts of the spectrum.

Light from atoms

*Like all other electromagnetic radiation, light emanates from atoms.
And the key to light's creation lies in the atom's structure.*

From the soft glow of a candle to the harsh glare of an arc lamp, all light comes from atoms. Scientists debated the nature of light for centuries – some arguing that it is waves like waves in water, others that it is streams of particles. Now they agree it shows qualities of both. Light is made up of little packets, or quanta, of energy called photons that behave en masse like waves. These photons are emitted by atoms.

Every time an atom emits a photon, it loses a little energy. Before electricity was tamed, most light came from processes that released this energy only by changing atoms or molecules irreversibly. The Sun's light, for instance, is generated when atoms fuse together in its intensely hot center. Candlelight comes from burning candle wax – an irreversible chemical reaction. Electric light bulbs, though, give light without changing irreversibly, because the atoms in the bulb's filament temporarily gain energy from the electric current.

How an atom's electrons are arranged and the energy they carry is the key to understanding light emission. Essentially, electrons orbiting near the nucleus have less energy than those orbiting far away. Indeed, each orbit has its own energy level – with energy levels rising in steps away from the nucleus. Electrons can be pumped farther out to a higher energy orbit by absorbing energy.

An atom with electrons pumped up to higher energy levels is said to be excited. Photons of light and other forms of radiation are emitted by an excited atom when electrons fall back to a lower energy orbit nearer the nucleus. The color of the light depends on how far the electron falls back, and on how much energy the photon contains. A photon of blue

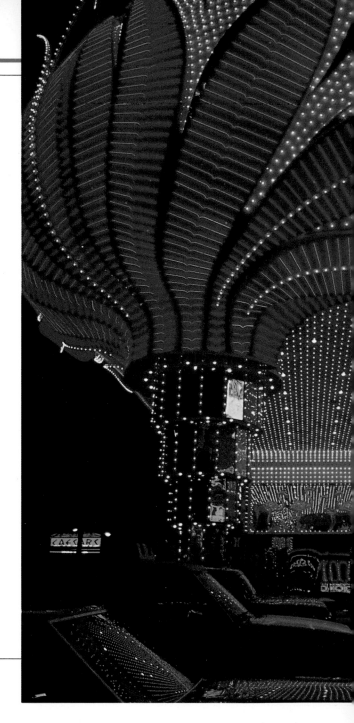

JUMPING FOR LIGHT

When an atom is hit by a photon – an energy packet of light – one of its electrons may be shifted from a low-energy state or orbit into a higher energy state (**right**). Almost immediately, the electron falls back; and in doing so, a photon is given off that has as much energy as the electron falling back has lost (**far right**).

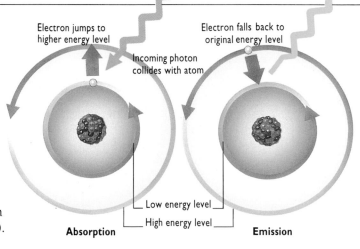

Photon given off

Electron jumps to higher energy level

Incoming photon collides with atom

Electron falls back to original energy level

Low energy level

High energy level

Absorption

Emission

A fluorescent light (right) is a discharge tube filled with vapor of the metal mercury at low pressure. Terminals at each end send electrons through the gas which hit the mercury atoms, exciting them and causing their electrons to jump to higher energy levels. When the electrons fall back to their former levels, they emit photons of ultraviolet light. These are invisible, but when they strike atoms in the phosphor coating the inside of the tube, they pump the electrons of the phosphor atoms to higher energy levels. On falling to their former levels, these emit photons as visible light of all colors, which mix to make white light.

The bright lights of Las Vegas depend, like all other light sources, on electrons changing energy levels in atoms. Most of the lights are discharge tubes – glass tubes containing low-pressure gas. When a high-voltage current is sent through the tube, it strips electrons off the gas particles. These electrons recombine with other gas particles and first excite the atoms, which then emit light as their electrons fall to a lower energy level. The color of the light depends on the gas. Neon gives red, sodium pale yellow, and krypton deeper yellow.

(shorter wavelength) light contains about twice as much energy, for instance, as a photon of red (longer wavelength) light. Heated metal glows first red, then yellow, then white as it gets hotter, because the greater the energy involved, the more shorter wavelength light is emitted. In addition to visible light, all the wavelengths of the electromagnetic spectrum, from radio waves to gamma rays, are emitted by electrons falling back in this way.

Television sets make use of this; phosphor dots of blue, red, and green create the picture on a color screen. The dots glow when their atoms are excited by a stream of electrons that scans across the back of the screen.

An electric stove burner glows red hot because its atoms are emitting photons. The electric current heats the atoms in the burner, and they begin to move very fast. As they zoom around, they bump into each other, exciting electrons to higher energy levels. As the electrons subside to their former energy levels, they send out photons of red light.

Fluorescent jackets glow in the dark because they absorb radiation we cannot see and re-emit it immediately in a form we can see. But they cannot be seen in a totally dark room because there is not even any invisible radiation to make them glow. A luminous watch, however, can glow in complete darkness because it absorbs energy when there is light around, then goes on re-emitting it slowly for quite a few hours afterward.

1 Electrode sends out continuous stream of electrons

Free electron

Electrode

Glass tube filled with mercury vapor

Inside of tube coated with phosphor

2 Electron collides with mercury atom causing electron to jump to higher energy level

3 Electron falls back to original orbit emitting ultraviolet photon

4 Atom in phosphor coating absorbs ultraviolet photon

5 Phosphor atom gives off visible photon

Key to atoms

Mercury atom

Excited mercury atom

Phosphor atom

Excited phosphor atom

Uranium 23

4.5 billion years

24 days

1 minute

Elements of instability

Some atoms are far from stable. They break up spontaneously, changing into different elements and giving rise to radioactivity.

The individual atoms of an element are not strictly identical. While each of an element's atoms has the same number of protons in its nucleus, some have a few neutrons more or less. This does not affect an atom's chemical behavior, but it makes the nucleus slightly heavier or lighter. For instance, along with 17 protons, some chlorine nuclei have 18 neutrons and others have 20. There are, in fact, three times as many lighter chlorine-18 nuclei as there are atoms of the heavier chlorine-20 nuclei.

These slightly heavier or lighter versions of an element are called isotopes; most elements have two or more such forms. Even hydrogen has isotopes. One in every 10,000 hydrogen nuclei has a neutron as well as a proton, doubling the atom's mass to make "heavy hydrogen" called deuterium.

Nuclei with equal numbers of neutrons and protons tend to be stable. But nuclei with an excess of neutrons or with many neutrons (especially over 82) tend to break up spontaneously.

The disintegration or decay of isotopes is called radioactivity, and the fragments created spread into the surroundings as invisible radiation. Radioactive decay is always going on around us to some degree. The decay of radium in granite rock, for instance, creates radioactive radon gas, which can, in a few areas, build up in homes and endanger health. But the highest levels of radioactivity are associated with nuclear reactions.

0 — Year 0
64 x C14 atoms

Half lives

1 — After 5,700 years
32 x C14 atoms

2 — After 11,400 years
16 x C14 atoms

3 — After 17,100 years
8 x C14 atoms

4 — After 22,800 years
4 x C14 atoms

5 — After 28,500 years
2 x C14 atoms

6 — After 34,200 years
1 x C14 atom

Carbon usually exists in two forms (below): stable, common carbon-12; and carbon-14, less common and less stable. Carbon-12 has six protons and six neutrons in its nucleus. An atom of carbon-14 also has six protons, but it has eight neutrons in its nucleus. This makes it unstable, and it tends to decay to nitrogen-14, giving off radioactivity in the form of a beta particle and gamma radiation.

C12
Stable

C14
Unstable

Nucleus contains 6 protons and 6 neutrons

Nucleus contains 6 protons and 8 neutrons

HALF-LIFE AND HISTORY

The decay of carbon-14 or radiocarbon has proved invaluable to scientists who use it to find out the age of anything containing organic matter – from human remains to old bits of wood. Carbon-14 and carbon-12 are present in a set ratio in all living things. When they die, the carbon-14 decays at a steady, unvarying rate. After 5,700 years, only half the original carbon-14 is left. Carbon-14 thus has a half-life of 5,700 years. After another 5,700 years (11,400 years in total), a quarter of the original amount is left. After three half-lives (17,100 years), only one-eighth remains. From a starting point of 64 carbon-14 atoms, after 34,200 years only one would remain (**left**).

Scientists work out when a thing died by seeing how many atoms of carbon-14 are left. Using carbon-14 dating, the year the man (**right**) found frozen in a glacier in the Alps met his death has been worked out to about 3000 B.C.

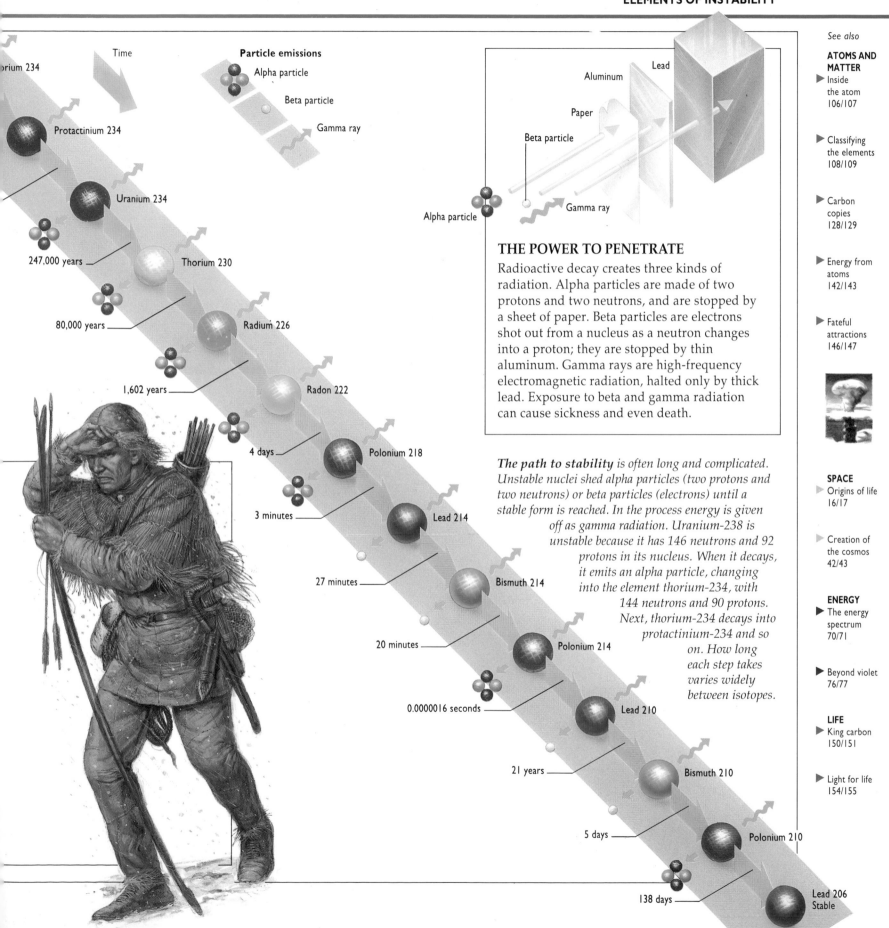

Time

Particle emissions

Alpha particle

Beta particle

Gamma ray

orium 234

Protactinium 234

Uranium 234

247,000 years

Thorium 230

80,000 years

Radium 226

1,602 years

Radon 222

4 days

Polonium 218

3 minutes

Lead 214

27 minutes

Bismuth 214

20 minutes

Polonium 214

0.0000016 seconds

Lead 210

21 years

Bismuth 210

5 days

Polonium 210

138 days

Lead 206
Stable

Alpha particle

Beta particle

Paper

Aluminum

Lead

Gamma ray

THE POWER TO PENETRATE

Radioactive decay creates three kinds of
radiation. Alpha particles are made of two
protons and two neutrons, and are stopped by
a sheet of paper. Beta particles are electrons
shot out from a nucleus as a neutron changes
into a proton; they are stopped by thin
aluminum. Gamma rays are high-frequency
electromagnetic radiation, halted only by thick
lead. Exposure to beta and gamma radiation
can cause sickness and even death.

*The path to stability is often long and complicated.
Unstable nuclei shed alpha particles (two protons and
two neutrons) or beta particles (electrons) until a
stable form is reached. In the process energy is given
off as gamma radiation. Uranium-238 is
unstable because it has 146 neutrons and 92
protons in its nucleus. When it decays,
it emits an alpha particle, changing
into the element thorium-234, with
144 neutrons and 90 protons.
Next, thorium-234 decays into
protactinium-234 and so
on. How long
each step takes
varies widely
between isotopes.*

Energy from atoms

Nuclear weapons and power stations unleash some of the phenomenal energy locked inside every atomic nucleus.

Nuclear fuel is so concentrated a form of energy that a couple of pounds or so of uranium yields as much energy as 400 barrels of oil – enough to produce a day's electricity for a city of 300,000 people. The energy comes from the forces that bind together the nuclei of atoms. Nuclear power stations get this energy by splitting atomic nuclei in a process called nuclear fission. Uranium is used because its huge nuclei are naturally unstable and easily split.

Nuclear power stations use the heat released by fission in a reactor to make steam to drive electricity-generating turbines. Several systems are used, including gas-cooled, water-cooled, and sodium-cooled reactors. Each has its pros and cons, but all generate hazardous radioactive waste which remains potentially dangerous for a long time. Because of nuclear fission's need for scarce uranium and the hazards from the waste it creates, many scientists think the future of nuclear energy lies with nuclear fusion (**opposite**).

Nuclear energy is also exploited to create devastatingly destructive weapons. Many nuclear weapons set off a fantastically rapid fission chain reaction by hurling plutonium particles together. This violent chain reaction splits so many nuclei so quickly that it releases a phenomenal amount of heat energy, creating an explosion many times greater than the most powerful conventional bomb. It also floods the area around the explosion with deadly radiation, leaving it contaminated for years.

To release energy, nuclear fission depends on setting up a chain reaction. This begins when a neutron is fired into the nucleus of an atom of uranium fuel, splitting it in two and unleashing huge amounts of energy. It also releases three neutrons which shoot away to collide with other uranium nuclei.

These nuclei, in turn, are split by the collision, releasing more heat and firing out more neutrons, and causing further fissions. Once started, the process goes on in this way until the fuel is used up. In an atomic bomb, the fuel is burned up within a tiny fraction of a second, unleashing huge destructive power. In a nuclear reactor, the process is

slowed down to make sure that the energy is released bit by bit.

A problem with naturally occurring uranium is that it contains several different isotopes. Only 7 atoms in 1,000 are uranium-235; most of the rest are uranium-238, which does not split as readily as uranium-235, and can inhibit the chain reaction. Some power stations get around this problem by enriching the fuel – that is, by boosting the uranium-235 content. Others slow the neutrons down by making them ricochet off atoms of a material such as graphite, called a moderator, and so raising their chances of hitting a uranium-235 atom.

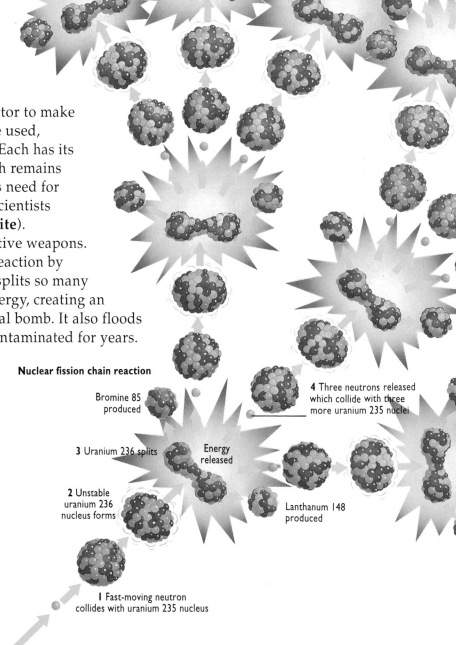

Chain reaction continues

Nuclear fission chain reaction

Bromine 85 produced

4 Three neutrons released which collide with three more uranium 235 nuclei

3 Uranium 236 splits

Energy released

2 Unstable uranium 236 nucleus forms

Lanthanum 148 produced

1 Fast-moving neutron collides with uranium 235 nucleus

POWER WITHOUT LIMIT

The Sun's heat comes from nuclear reactions in its center where high pressure and temperature push hydrogen nuclei together in a process called fusion. Scientists are trying to make fusion a practical reality. If successful, it would provide plentiful power at a low cost, since the raw materials are found in sea water and the products appear to be harmless. The largest fusion experiment which uses magnetic fields to control the fuel is JET (Joint European Torus) at Culham, England. In JET abundant fusion reactions are achieved, and for the first time in a fusion device, nearly 2 MW of power was produced in 1991.

The Joint European Torus (JET) (**below**) *is 39 feet (12 m) and 49 feet (15 m) across. At its heart is the chamber (**bottom**) where nuclear fusion takes place at colossally high temperatures of up to 540 million °F (300 million °C).*

Nuclear fusion

Heat

Helium produced

Energy released

Neutron released

Deuterium

Pressure

Tritium

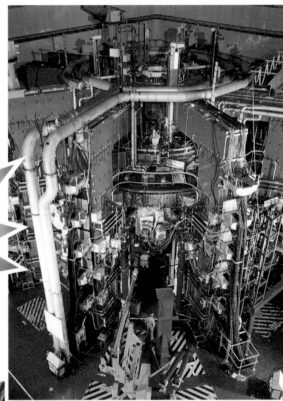

*Most research projects aim to achieve fusion by mimicking the conditions in the heart of the Sun. Inside the JET fusion reactor chamber, a plasma – a very hot gas – is held within a torus, which is a huge doughnut-shaped vacuum vessel (**right**). Magnetic fields hold the plasma away from the walls of the torus and the plasma is heated to extremely high temperatures. In the plasma, nuclei of the heavy isotopes of hydrogen – deuterium and tritium – are fused together. When this happens, much energy is given off, and helium and neutrons are formed (**above**).*

Exotic particles

Modern science has revealed a fascinating and complicated subatomic world teaming with all kinds of particles – some with strange names.

Scientists once believed that atoms were simple structures made up of just three fundamental particles – electrons, protons, and neutrons. But studies of radioactivity showed that when a neutron decayed into a proton releasing an electron (beta decay), there was a tiny bit of mass not accounted for. This could be explained only by the existence of a ghostly particle called a neutrino. Then special particles, called muons, like heavy electrons, and pions or pi-mesons, which bind protons and neutrons together, were discovered in cosmic rays. Soon accelerators built to smash nuclei to bits by hurling particles together were revealing scores more.

Eventually, physicists were confronted with a bewildering zoo of subatomic particles – until they realized all might be grouped into three families. Electrons, muons, and neutrinos, for instance, are all variations on the same particle, called a lepton; while protons, neutrons, and pions are all kinds of hadron. A third family, called bosons, includes tiny messenger particles that transmit all the basic forces of the universe. Photons, for instance, are the bosons that carry electromagnetic force, and there may be particles called gravitons responsible for gravity.

Now physicists believe all hadrons are made up from even more basic particles, called quarks. According to the quark theory, quarks come in six different "flavors" – up, down, strange, charmed, top, and bottom. Neutrons and protons are basically triplets of quarks; pions are pairs. Along with leptons, quarks seem to be the fundamental building blocks of the universe.

As if this were not enough, physicists have long been convinced that each and every particle has an antiparticle, its invisible mirror image – alike, but opposite in every way. For every electron, there is an invisible positively charged positron. For every quark there is an antiquark. And so on. It was once thought there might be as much antimatter in the universe as matter, but now scientists believe nearly all of it was destroyed soon after the Big Bang, along with most matter – leaving behind the small quantity of matter in the universe today.

Many new subatomic particles have been spotted in special detection equipment including cloud chambers and bubble chambers by the trails left behind after particles collide at high speed. Cloud chambers are sealed glass vessels containing moist air. Charged subatomic particles leave a detectable trail of condensed moisture in a cloud chamber like the vapor trail behind a jet aircraft.

The image from a bubble chamber containing very cold liquid hydrogen (key, right and far right) shows just some of the many particles produced by high-speed collisions of tiny particles. A negative kaon (yellow) enters from below and collides with a proton in the liquid hydrogen filling the chamber. It produces a positive pion (salmon pink) and a negative pion (purple) plus a neutral lambda particle which is not visible in this image. The lambda is revealed when it decays into a proton (red) and a muon (pale blue).

The muon soon decays into an electron (the larger green spiral) and an invisible neutrino. The positive pion (pink) knocks an electron off an atom which forms the tighter of the two spirals.

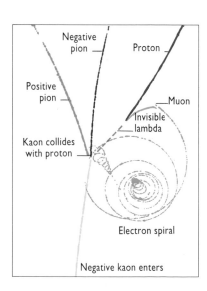

Negative pion — Proton

Positive pion —

— Muon

Invisible lambda

Kaon collides with proton

Electron spiral

Negative kaon enters

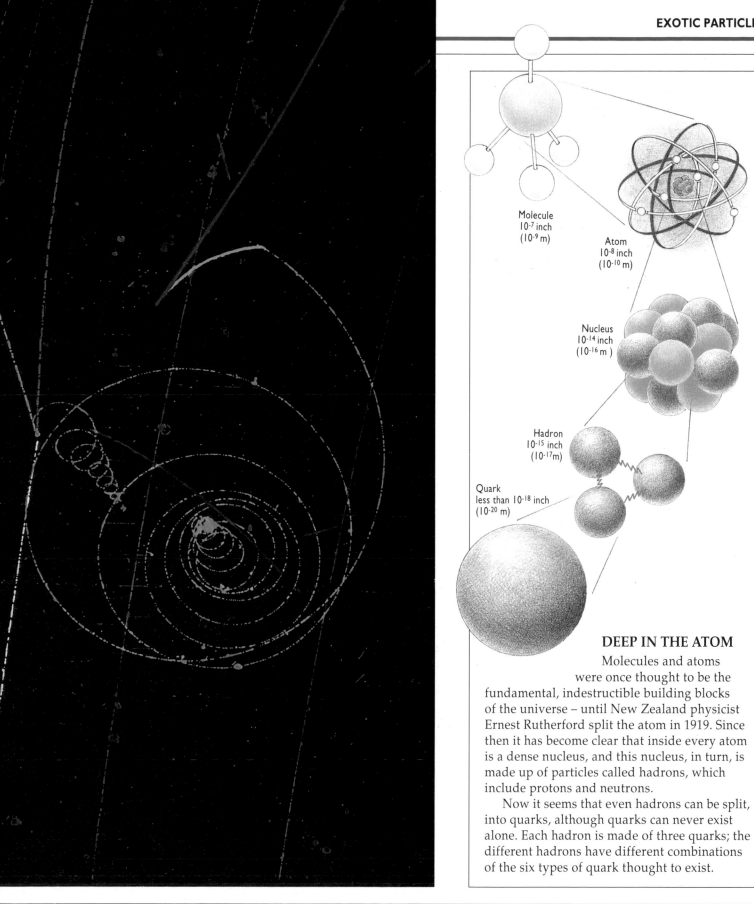

Molecule
10^{-7} inch
(10^{-9} m)

Atom
10^{-8} inch
(10^{-10} m)

Nucleus
10^{-14} inch
(10^{-16} m)

Hadron
10^{-15} inch
(10^{-17} m)

Quark
less than 10^{-18} inch
(10^{-20} m)

DEEP IN THE ATOM

Molecules and atoms
were once thought to be the
fundamental, indestructible building blocks
of the universe – until New Zealand physicist
Ernest Rutherford split the atom in 1919. Since
then it has become clear that inside every atom
is a dense nucleus, and this nucleus, in turn, is
made up of particles called hadrons, which
include protons and neutrons.

Now it seems that even hadrons can be split,
into quarks, although quarks can never exist
alone. Each hadron is made of three quarks; the
different hadrons have different combinations
of the six types of quark thought to exist.

Fateful attractions

All matter is held together – or sometimes even pushed apart – by just four fundamental forces.

Scientists believe that all matter in the universe is subject to just four basic forces – gravity, electromagnetism, and the strong and weak nuclear forces. The strong nuclear force binds protons and neutrons together in the atomic nucleus, despite the electrical repulsion between the positively charged protons which otherwise would force them apart. The weak nuclear force – a tiny fraction (10^{-23}) of the strength of the strong nuclear force – is the force that breaks up neutrons, forming a proton and expelling an electron as beta radioactivity.

The understanding of these four basic forces has been profoundly influenced by quantum theory, the idea that radiation is emitted not in continuous waves, but in little packets of energy called quanta. The idea was first applied to electromagnetic fields, and quantum electrodynamics (QED) theory views all electromagnetic radiation in terms of quanta called photons, which are absorbed and emitted by electrons.

Each force seems to have its very own messenger particle like the photon. The weak force is carried by W and Z particles, the strong force by pions. Gravity may be transmitted by gravitons, as yet undetected.

With the discovery of quarks, the particles that make up protons and neutrons, the theory known as quantum chromodynamics (QCD) was developed to explain the forces between them.

In QED there is just one kind of messenger particle, the photon, but in QCD there are no fewer than eight, called gluons, which transmit the force and "glue" quarks together.

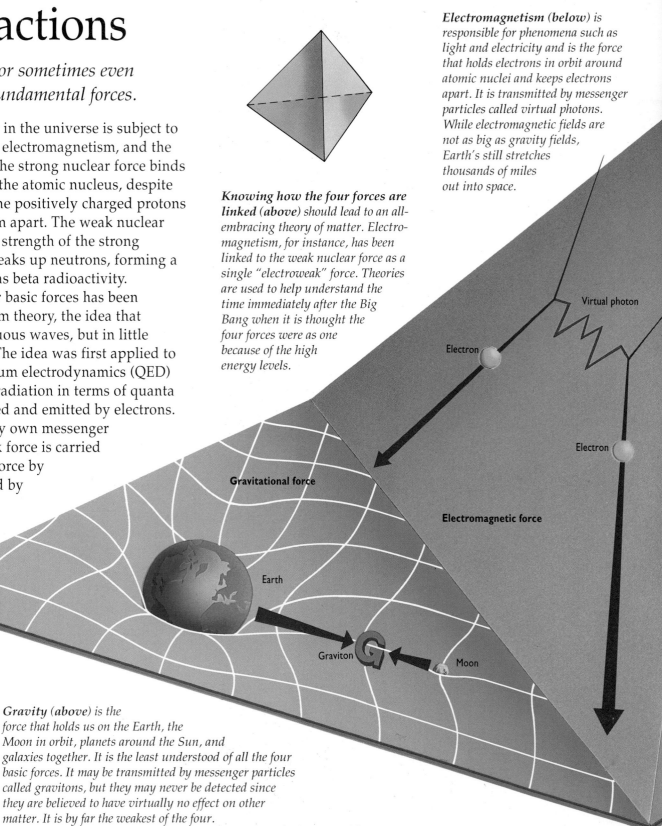

Knowing how the four forces are linked (above) should lead to an all-embracing theory of matter. Electro-magnetism, for instance, has been linked to the weak nuclear force as a single "electroweak" force. Theories are used to help understand the time immediately after the Big Bang when it is thought the four forces were as one because of the high energy levels.

Electromagnetism (below) is responsible for phenomena such as light and electricity and is the force that holds electrons in orbit around atomic nuclei and keeps electrons apart. It is transmitted by messenger particles called virtual photons. While electromagnetic fields are not as big as gravity fields, Earth's still stretches thousands of miles out into space.

Virtual photon

Electron

Electron

Gravitational force

Electromagnetic force

Earth

Graviton

Moon

Gravity (above) is the force that holds us on the Earth, the Moon in orbit, planets around the Sun, and galaxies together. It is the least understood of all the four basic forces. It may be transmitted by messenger particles called gravitons, but they may never be detected since they are believed to have virtually no effect on other matter. It is by far the weakest of the four.

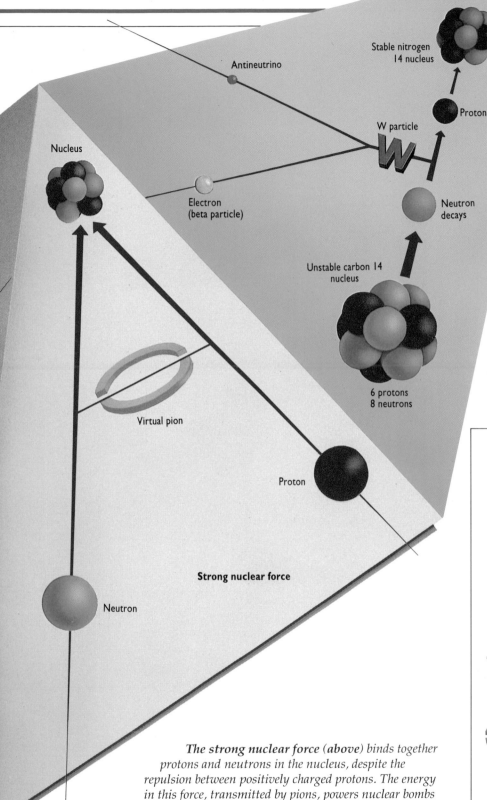

Weak nuclear force

Stable nitrogen
14 nucleus

7 protons
7 neutrons

Antineutrino

Proton

W particle

Nucleus

Electron
(beta particle)

Neutron
decays

Unstable carbon 14
nucleus

6 protons
8 neutrons

Virtual pion

Proton

Strong nuclear force

Neutron

The strong nuclear force (above) binds together protons and neutrons in the nucleus, despite the repulsion between positively charged protons. The energy in this force, transmitted by pions, powers nuclear bombs and keeps stars burning. It is effective over a short distance, so it can only bind smaller nuclei powerfully. Big nuclei are less strongly bound, and so unstable, because of the extra repulsion between large numbers of protons.

The weak nuclear force (left) is the force that splits neutrons, causing beta radioactivity. It is carried by W and Z particles in much the same way as an electric current. Because W and Z particles are very large, the weak force is effective only over an extremely short distance.

When a neutron in an atomic nucleus decays into a proton and an electron (a beta particle), a W particle passes energy between the particles. The beta particle is emitted from the nucleus of the atom in which the neutron is changing. Also given off in beta decay is an antineutrino, a mysterious massless – yet energy-containing – particle that has no charge.

FOUR-FORCE LINE-UP

The four forces act over different distances and have widely differing strengths. In fact, the difference between the strongest, the strong nuclear force, and the weakest, gravity, is 38 orders of magnitude. Thus gravity is 10^{38}, that is 10 with 37 zeros after it, times weaker.

G
Graviton

Virtual photon

Virtual pion

W
W particle

Gravity, though a feeble force, acts over infinite range. Gravitons would be the smallest messenger particles. The electromagnetic force is much stronger, and it, too, has infinite range. Its messenger particle is the virtual photon, and it is 10^{36} times as strong as gravity.

The strong force, whose messenger particle is the virtual pion, has a small range of about 10^{-12} mm, and compared with gravity, it is 10^{38} times as strong. The weak nuclear force has W and Z messenger particles and is approximately 10^{15} times as strong as gravity. Its range is minute at only 10^{-15} mm.

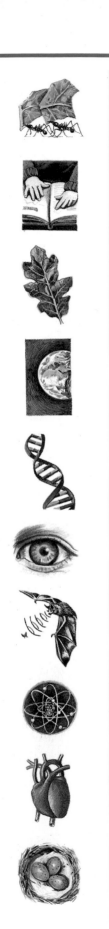
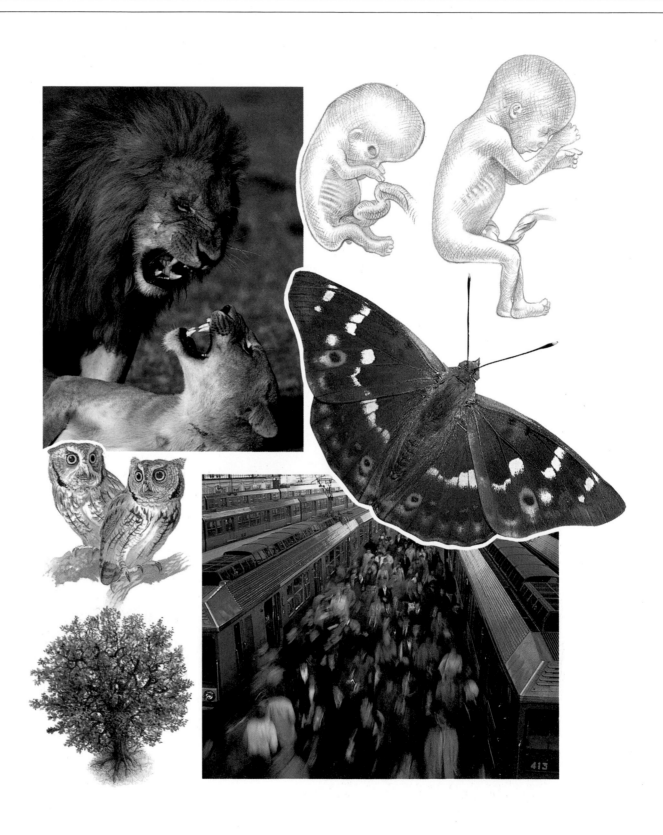

Life

F or nearly 4 billion years, life has existed on Earth. In that time, it has evolved into a multiplicity of forms that have yet to be fully cataloged and understood. The ways organisms work, including their internal processes, are as complex as their external relationships with the world and the other living things in it.

Despite the almost bewildering abundance and diversity of life on Earth, all plants and animals obey some simple rules. They harness energy, they process raw materials, they grow, and they respond to the world around them. How efficiently they carry out these functions helps to determine whether they succeed or perish. But their survival depends on one paramount ability, shared by all living things – the ability to reproduce. Reproduction depends on complex carbon-based molecules – the mysteries of which scientists are now starting to unravel.

*Left clockwise from top: lions mate; a fetus develops; a butterfly warms itself in the sun; living together in cities; a tree is constructed with carbon; night-hunting owls. **This page (top):** the structure of a cell; (**left**) a flower that follows light.*

King carbon

Everything living, from a leaf on a tree to a germ in a test tube, is made with the element of life – carbon.

All living things are like immensely complicated construction sites, in which atoms are the building blocks. There are about 35 different elements essential to the constant building, demolition, and recycling of materials of life. But by far the most important is the atomic cornerstone of living structures – carbon, the element that also forms diamonds and graphite.

Carbon atoms have the unique ability to link up with either other carbon atoms or atoms of different elements to form chains and clusters that are unusually stable and strong. These chains can contain just a handful of atoms, or many millions. Each different arrangement of the atoms produces a different carbon compound with its own unique chemical characteristics.

Some carbon compounds are strong and may form cell walls or tendons. Others, such as the carbohydrates sugar and starch, store energy from the Sun when their chemical bonds are built up, and when their bonds are broken down it is released. Waxy carbon compounds, called lipids, repel water and are used to seal off living matter from watery surroundings. Carbon-based enzymes dramatically speed up chemical reactions. Some carbon compounds even act as data banks, holding chemical information, and one – DNA – the master data bank itself, passes on information from one generation to the next.

Carbon, the most important element of life, plus three other essentials, oxygen, hydrogen, and nitrogen, take part in immense global cycles that link the living world with the non-living one. A carbon atom that today exists in a piece of rock may once have formed part of a bone or a leaf, and in the future, may once again find its way into living matter.

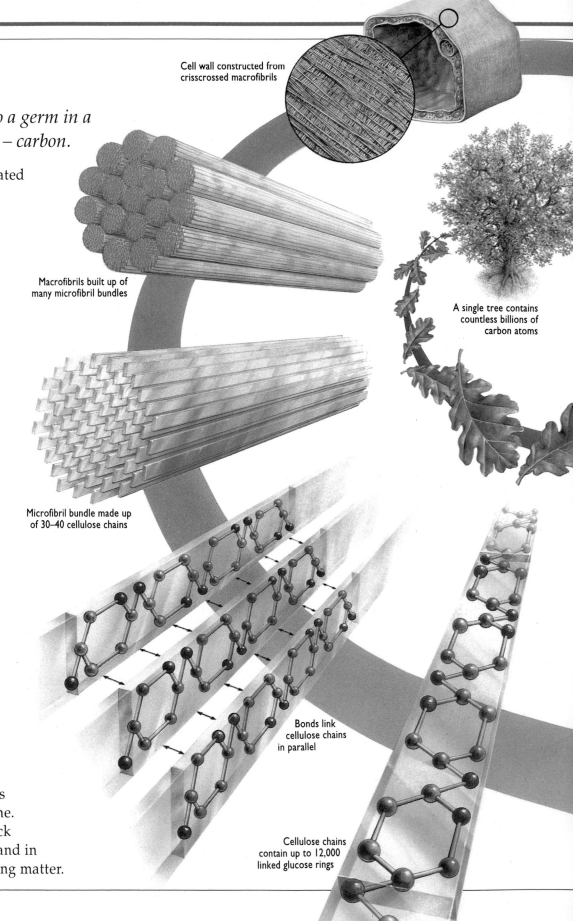

Cell wall constructed from crisscrossed macrofibrils

Macrofibrils built up of many microfibril bundles

A single tree contains countless billions of carbon atoms

Microfibril bundle made up of 30–40 cellulose chains

Bonds link cellulose chains in parallel

Cellulose chains contain up to 12,000 linked glucose rings

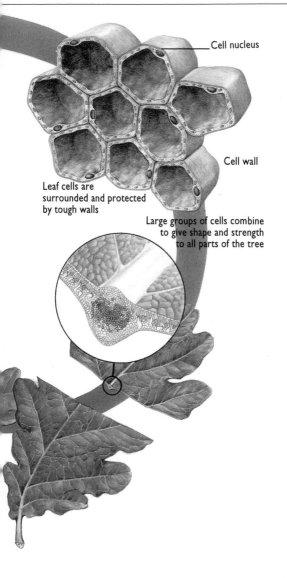

Cell nucleus

Cell wall

Leaf cells are surrounded and protected by tough walls

Large groups of cells combine to give shape and strength to all parts of the tree

A tree is a living thing, so it is assembled from carbon structures which are, when looked at in greater detail, made of smaller and smaller carbon-built components.

The shape and strength of every leaf depends on the arrangement of the leaf's cells. Each of the leaf's thousands of cells has tough walls, fortified with a multi-layered, crisscross arrangement of tough stringy fibers called macrofibrils. Individual macrofibrils are formed from bundles of even smaller fibers, called microfibrils. Each microfibril is a long chain of cellulose, the most abundant of the many carbon compounds made by living things. The attractions between hydrogen atoms of neighboring cellulose chains hold the microfibrils together.

Each chain of cellulose is composed of many thousands of glucose molecules joined together. However, cellulose has one big advantage over its constituent glucose molecules. It does not dissolve or disintegrate in water, and this makes cellulose the perfect base for lasting, strong structures like leaves, trunks, and branches.

SILICON: PRETENDER TO THE THRONE

Many science-fiction stories have been written about forms of life based not on carbon, but on some other element. Silicon is a particularly popular candidate. Found in sand and many kinds of rock, silicon is one of Earth's most abundant elements. Carbon and silicon are in the same chemical group, so they are similar in many ways. Like carbon, every silicon atom can form four chemical bonds and can link together to make long chains. However, silicon compounds are far less stable than their carbon equivalents, and in water large silicon molecules tend to disintegrate. Because water is vital to life on Earth, this instability makes it impossible for silicon atoms to build the self-copying chemicals which are the basis of life.

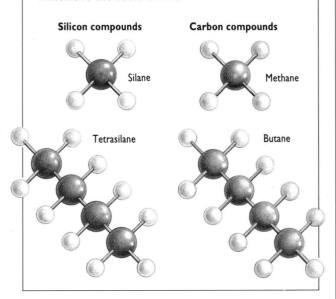

Silicon compounds

Silane

Tetrasilane

Carbon compounds

Methane

Butane

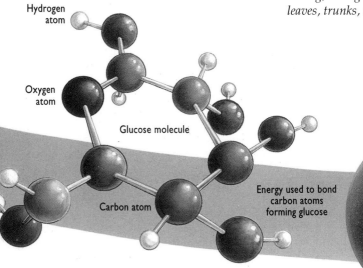

Hydrogen atom

Oxygen atom

Glucose molecule

Carbon atom

Energy used to bond carbon atoms forming glucose

Carbon atom

The ring of carbon atoms at the center of every glucose molecule makes the process of plant-building possible. Strings of glucose molecules join together to make cellulose, the major structural component of plants. Glucose, a type of sugar, is one of the basic stores of energy in the food chain. It is made by the process of photosynthesis using carbon dioxide from the air and water from the soil. During photosynthesis, a plant's leaves absorb energy from the Sun and use it to create the chemical bonds that join atoms of carbon with those of oxygen and hydrogen to form molecules of glucose.

See also

LIFE
▶ Order from energy
152/153

▶ Light for life
154/155

▶ Blueprint for life
170/171

▶ The microscopic factory
174/175

SPACE
▶ Origins of life
16/17

ENERGY
▶ Earth's energy store
60/61

ATOMS AND MATTER
▶ The key to behavior
106/107

▶ Classifying the elements
108/109

▶ The intimate bonds
112/113

▶ Carbon copies
128/129

▶ Carbon's vital links
130/131

Order from energy

Snatching temporary organization from the chaos of the universe,
living things use energy to create structure and order.

Leave a flashlight on, and eventually the bulb dims before it finally goes out. As the battery gets flat, its electricity-generating chemicals mix up and become disordered. It is the same in the universe as a whole. The process of running down and becoming disordered is an inescapable law of physics, and energy given out along the way eventually turns to diffuse background heat.

But on Earth – and perhaps in other places, too – there are "islands" where, on a purely local scale, the process is reversed. These islands are living things that take in energy and use it to create order at the expense of their surroundings. The results of this process can be astonishing and beautiful. A snail's shell, a peacock's feather, a whale's brain, and a plant's cell are all highly ordered systems, made by living organisms. So, too, are the structures living things build, such as webs and nests. All through the natural world, order is being temporarily formed from chaos.

Living things obtain energy for making order in two ways. Some organisms, known as autotrophs, collect energy directly. The usual source of this energy is sunlight, and foremost among these collectors of sunlight are plants, which use photosynthesis. Heterotrophs get their energy indirectly, by feeding on autotrophs and digesting the carbon-based, energy-containing compounds they are made up of. Humans are heterotrophs, as are all other members of the animal kingdom. When animal eats plant, and then animal eats animal, the energy that came from the Sun is passed through a chain of living things.

The wax combs inside a beehive (left) vividly show how living things can organize matter. The cells in a comb are used for raising young bees and for storing honey. They are built with almost mathematical accuracy, to give the maximum strength and storage space with the minimum use of materials. Bees do not have to learn how to construct these beautifully designed structures; they are guided entirely by instinct.

The highly ordered cells from a nettle's stem (right) are built using energy collected from sunlight. A single plant can contain millions of cells, and each plays a part in sustaining life and producing the next generation. The plant creates all this complexity from the simplest raw materials – water, carbon dioxide, minerals, and energy.

ALIVE OR DEAD?

Viruses inhabit a strange twilight world at the boundary between living and non-living things. If an organism is alive, then it is able to reproduce itself. Viruses obey this rule, but in a rather underhand way. They can multiply themselves only by infecting living cells and then commandeering the cells' chemical machinery so that they abandon ordinary work and instead make copies of the virus.

But outside living cells, viruses show no signs of life. They cannot harness energy by themselves, which means they cannot react to the world around them or reproduce without pulling off a hijack.

Viruses, such as the Adenoviruses (**below**), which cause symptoms similar to the common cold, are much smaller than ordinary cells. They are nothing more than a tiny amount of genetic material housed inside a casing made of protein. A virus's chemical structure is so precise that, when dried, viruses can even be made to form crystals. If the crystals are then dissolved in water, the virus particles can come alive by infecting a host.

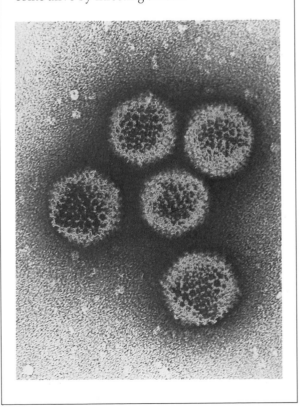

Light for life

*Green plants are the key to all life on Earth because they capture the
Sun's energy using the process called photosynthesis.*

Like thousands of living tiles, a tree's leaves overlap to form a roof of greenery that
intercepts much of the Sun's light. But trees do not do this simply to act as a
sunshade. Instead, like all green plants, they use their leaves to capture energy.
When sunlight hits a leaf, its energy is trapped by a substance called chlorophyll –
the chemical that makes leaves green. Once chlorophyll has captured the
Sun's energy, it passes this on through a series of chemical reactions known
as photosynthesis.

Photosynthesis means "building up by light." Plants use carbon dioxide
and water as their raw materials. They get carbon dioxide from the air with
their leaves, and water from the ground with their roots. Using the Sun's
energy, they combine them, first forming glucose, then transforming glucose into
a variety of more complex substances, such as starch and cellulose. Oxygen, a
waste product of photosynthesis, is released into the atmosphere.

For animals, this oxygen is a crucial by-product. Photosynthesis first
began about 3 billion years ago, when it was carried out by bacteria.
These bacteria combined carbon compounds with substances such as
hydrogen sulfide, producing sulfur as waste. But when bacteria
became able to use water in photosynthesis, a profound change
gradually took place over the whole Earth, because for the first
time, free oxygen entered the atmosphere. Thanks to photosynthetic
bacteria and later green plants, the air became rich in the gas,
allowing oxygen-breathing animals to spread over the Earth.

Plants capture only about one-ten
thousandth of the solar energy that
falls on Earth, but this energy is the
ultimate driving force behind nearly
all forms of life on our planet.

The substance that collects the
energy – chlorophyll – is not
scattered throughout the cells that
make up leaves. Instead, it is found
inside disk-shaped structures called
chloroplasts. The rest of a leaf cell is
colorless, so light that does not meet
a chloroplast passes through.

Within the chloroplast, which is
bounded by outer and inner
membranes, are smaller structures.
The active region of the chloroplast
is called the thylakoid. Here, energy

from chlorophyll "fixes" carbon
from carbon dioxide in the
atmosphere, building it into a
form that the plant can use.
The reaction uses six
molecules each of water and
carbon dioxide, and makes
them into one molecule of
glucose and six of oxygen.

Plants use a variety of
carbon-containing
substances to store
the captured energy.
In a potato, starch is
the chemical used for
storage. It is made
from many glucose
molecules.

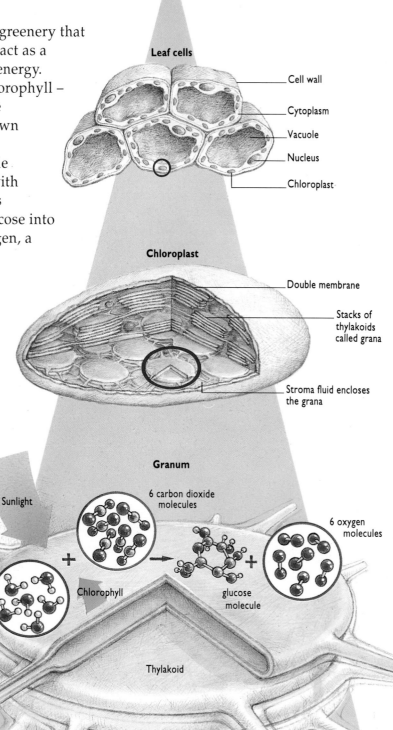

Leaf cells

Cell wall

Cytoplasm

Vacuole

Nucleus

Chloroplast

Chloroplast

Double membrane

Stacks of
thylakoids
called grana

Stroma fluid encloses
the grana

Granum

Sunlight

6 carbon dioxide
molecules

6 oxygen
molecules

6 water
molecules

Chlorophyll

glucose
molecule

Thylakoid

Lush and green, the plants of the forest stretch up to reach for the available sunlight. Without energy from the Sun, life on Earth would cease to exist – probably.

In 1979, scientists stumbled across a strange community of living things over 1½ miles (2.5 km) deep in the Pacific Ocean. Around vents of volcanically heated water were crabs, mussels, and giant tube-dwelling worms. Active vents abounded with life, while those whose water had ceased to flow were surrounded by dead remains.

Biologists have found that the worms live in partnership with bacteria that process chemicals in the mineral-rich water, and the crabs and mussels live on their remains. The vent-dwellers are the only animals on Earth that do not depend for survival on photosynthesis powered by the Sun.

Photosynthesis and respiration are two complementary processes, with the waste products from one forming the raw materials for the other (*below*). The two create an interlocking cycle, which maintains the levels of oxygen and carbon dioxide in the atmosphere.

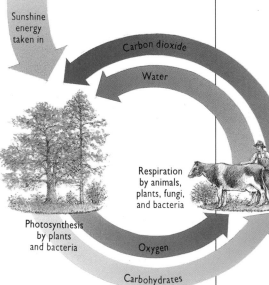

Sunshine energy taken in

Carbon dioxide

Water

Respiration by animals, plants, fungi, and bacteria

Photosynthesis by plants and bacteria

Oxygen

Carbohydrates

Keeping your balance

Living things rely on constant feedback of information about changing conditions to keep them working smoothly.

When you switch on an oven, a thermostat makes sure that once it has warmed up, it stays at the temperature you have selected. The thermostat does this by detecting changes in temperature. If the oven gets too hot, it reduces the electricity or gas supply; if the oven gets too cold, it increases it. The result is that the oven's temperature stays within narrow margins.

An oven is an example of a negative feedback system, one in which any change from a pre-set condition tends to cancel itself out. Negative feedback is used in many manmade devices, from thermostats to automatic pilots. But as complex as many of these are, they do not begin to rival the most advanced feedback systems of all – those found in living things.

If it is to work properly, a cell has to maintain a precise internal environment. It has to monitor and regulate salt concentrations, water balance, and many other factors. This internal balancing act is known as homeostasis. If the balance is not properly maintained – if, for example, the cell takes in too much water – its shape or its chemical processes will be disrupted. If the disruption is severe, it may die.

All forms of life use feedback to make sure that this does not happen. The human body is made of many cells, and particular groups of cells carry out the task of sensing changes and making appropriate adjustments. For example, the body's thermostat lies in an area of the brain known as the hypothalamus. It detects any change in body temperature and quickly acts to correct it. The same

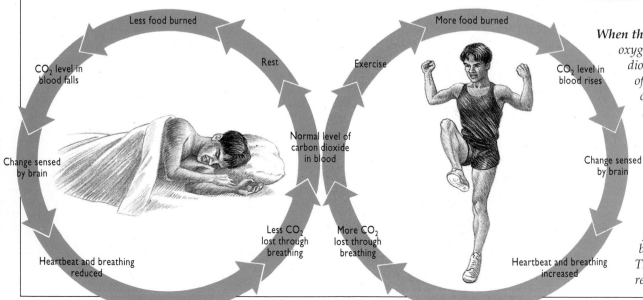

Less food burned • Rest • More food burned

CO_2 level in blood falls

Exercise

CO_2 level in blood rises

Change sensed by brain

Normal level of carbon dioxide in blood

Change sensed by brain

Heartbeat and breathing reduced

Less CO_2 lost through breathing

More CO_2 lost through breathing

Heartbeat and breathing increased

Rest

Exercise

When the body muscles work hard, they use up oxygen in the blood and replace it with carbon dioxide. They also generate heat. Cells in a region of the brain called the hypothalamus detect these changes and trigger responses to adjust them. As a result, the skin begins to produce sweat, which cools the skin as it evaporates. The heart beats faster, and the skin flushes as more blood is sent to it, carrying heat away from the muscles. At the same time, the breathing rate increases, so that more oxygen can be absorbed through the lungs, and more carbon dioxide lost. Even after exercise is finished, these changes continue until the body's normal balance has finally been restored. The opposite happens at rest, when breathing is reduced to match the low level of activity.

part of the brain also monitors food supply and water levels. If you go without food or water for any significant length of time, your hypothalamus issues the signals that you feel as hunger and thirst. By responding to them, your body is brought back into balance.

Quite often, feedback mechanisms in the body work through chemical messengers called hormones. The body's energy levels, for instance, are controlled by hormones released from the thyroid gland. If energy levels in the body are low, extra thyroid hormones are released, stimulating activity.

Not all living things have their controls pre-set in the same way. Mammals and birds are "warm-blooded," meaning that their bodies are set to maintain a particular warm temperature, regardless of the temperature all around. Most other animals are "cold blooded." Their temperature is more affected by their surroundings, and can rise and fall considerably in the course of a day. Even so, these animals still have their limits. If a snake or a lizard starts to get too hot, its inbuilt heat-sensing system tells it to take appropriate action, and it makes for the shade.

STAYING COOL

Many cold-blooded animals can control their temperature by their behavior. Reptiles such as lizards spend hours basking in the sun, soaking up its warmth. Once they are warm enough, they scurry off to find food, rest in the shade, then bask again.

Insects behave in similar ways. To warm up, a butterfly (**right**) opens its wings wide and faces them toward the sun. When its body has reached a perfect temperature, it brings its wings closer together. Some moths have taken temperature regulation a step farther. They "shiver" to generate heat and hold the heat with a mat of furlike tufts that acts as insulation.

Messengers of life

From the slow changes of puberty to the rapid reaction to a fright, hormones are essential to the body.

For an animal's body to work, the cells in one part often have to be able to send information to those in another. If you accidentally touch something hot, for example, a message has to travel from your fingers to the muscles in your arm, so that the hand is pulled away from the source of the heat.

In animals, including humans, urgent messages like this are flashed from one place to another as electrical impulses, which travel along nerves. But the body's cells can also communicate in a different way, by using special chemicals known as hormones.

If the nervous system is thought of as a telephone service, then the hormonal system is a postal equivalent. It usually works more slowly, but its effects can be far-reaching. The "letters" in this system are dozens of different hormones, specific chemicals that are manufactured in certain areas of the body, known as endocrine glands. When a hormone is released, it travels through the bloodstream until it is delivered to the correct address – its target cells. Here, the "letter" is read, and its instructions are acted on.

Hormones trigger all kinds of responses in the body, and many of them are involved in complex feedback mechanisms that keep the body in a steady state. Some hormones provoke the release of several

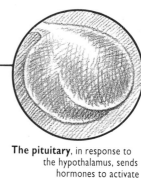

The hypothalamus, deep within the brain, regulates and monitors the whole hormone system.

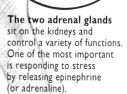

The pituitary, in response to the hypothalamus, sends hormones to activate other glands around the body.

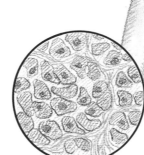

The thyroid gland, located in the front of the neck, controls the body's metabolism, influencing growth and development.

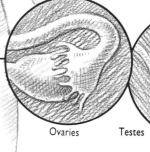

The two adrenal glands sit on the kidneys and control a variety of functions. One of the most important is responding to stress by releasing epinephrine (or adrenaline).

Cells in the pancreas produce insulin which regulates the level of sugar in the blood.

Ovaries   Testes

THE STRESS RESPONSE

When something makes you jump, fast-acting adrenaline is at work. In response to fear, anger, or stress, the body gets ready by releasing the chemical adrenaline, which has a profound effect on many parts of the body. Adrenaline prepares the body for emergency action by stepping up all the processes that might be involved in "fight or flight." In effect, it prepares you to run for your life or fight for it.

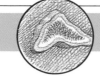

Reacting to fear or stress, **adrenal glands** dump adrenaline into the blood. The hormone quickly spreads and has its effects.

The **pupils of the eyes** dilate (get wider) so that more light enters and details of movement can be picked out more clearly.

Blood is diverted away from the skin (a non-essential area). This explains why people turn paler or look gray when they are afraid.

Adrenaline raises the **heart rate** and blood pressure so that the muscles are well supplied with blood if they are called on for action.

Adrenaline increases the **breathing rate** and expands the lung's air-absorbing sacs, boosting oxygen levels in the blood.

Many people actually seek out apparently dangerous situations, such as a ride on a roller coaster (**right**). Excitement, it seems, is good for us in manageable doses. An effect of getting a fright – even for fun – is to fire up the body's hormonal system and trigger the release of adrenaline.

other hormones, and many hormones work in pairs, with each one counteracting the effects of the other. By adjusting the output of each hormone, the body keeps in balance.

One fast-acting hormone is adrenaline. It primes the body to face danger by getting it ready to run or fight. Compared with adrenaline, most other hormones are much slower in their effects. Sex hormones, for example, prepare the body for reproduction, a process that in humans takes several years.

Plants do not have nerves, primarily because they do not move about and so do not need to react so quickly to the world around them. However, plants do have hormones. Most plant hormones are concerned with triggering kinds of growth and development. Hormones called auxins bend plants toward the light by speeding up growth on the side of a stem that is in the shade. Another plant hormone, called abscisic acid, prevents growth. It is this hormone that makes plants shed their leaves in the fall.

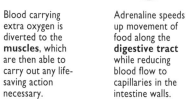

The ovaries and testes, in women and men, respectively, produce the sex hormones which control fertility and sexual characteristics.

Blood carrying extra oxygen is diverted to the **muscles,** which are then able to carry out any life-saving action necessary.

Adrenaline speeds up movement of food along the **digestive tract** while reducing blood flow to capillaries in the intestine walls.

To send more of the body's fuel to the muscles adrenaline makes the **liver** release more of its stored sugar into the blood.

Using light

Most plants and animals use light to regulate their lives – light dictates when plants grow and flower, and when birds and animals sleep, eat, mate, and migrate.

Many plants and animals run their lives to a strict timetable. Although they do not have alarm clocks or calendars, plants seem to know exactly when to flower and animals when to breed. One way they do this is by responding to light.

Every day, as the Sun rises and sets, it provides a cue for some animals to become active and for others to go to sleep. Many animals rely on the cover of darkness to protect themselves from their enemies, and they hide away during daylight. Others need light to find their food, and for them dawn is the signal to begin the daily search.

Plants are also sensitive to the daily progress of the Sun. Some turn their leaves and flowers to catch the maximum amount of sunshine. Many, such as dandelions, open their flowers at dawn and close them again at dusk, and some, such as gentians, are so sensitive to light that a passing cloud can make them close up.

At the equator, days are almost always the same length – 12 hours day and 12 hours night. Away from the Equator, however, day length varies markedly through the year, which gives us the seasons. This variation regulates all aspects of life for some creatures; for example, the colorado beetle can only grow and

The sunflower *is one of the natural world's most dedicated sunbathers. Throughout the day, the sunflower turns its head to follow the Sun's progress across the sky. Known as phototropism, this light-following behavior means that each plant receives the maximum amount of sunshine. The bending and turning of the flower is* controlled by the distribution of light-sensitive plant hormones. These hormones respond to the light intensity by concentrating on the shady side of the stem. The hormones trigger the elongation of the cells of the stem, which twists the flower toward the Sun.

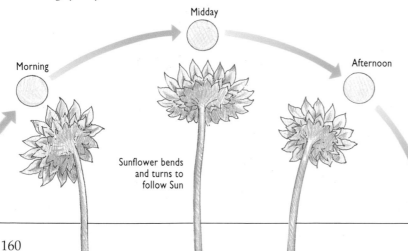

Morning

Midday

Afternoon

Sunflower bends
and turns to
follow Sun

reproduce in summer when there are at least 14 hours of light.

The shortening days of fall provide the signal for many trees to drop their leaves, for birds to migrate to warmer climes, and for some animals, like the hedgehog, to hibernate, or sleep through the cold winter months. The arrival of spring is heralded by warmer, longer days. Plants grow quickly and flower, birds arrive back from their winter retreats to build nests and lay eggs, and animals mate and rear their young.

Light is, of course, primarily used for seeing. Animals detect light through their eyes, which house special receptor cells that are sensitive to light energy. Humans and other mammals see with only a small part of the energy spectrum – visible light. But some animals see with other areas of the spectrum. Snakes such as pit vipers can see in infrared – the energy given off by warm objects. Using special sensory pits, they can detect the precise position of prey in total darkness.

Deep underwater, where sunlight does not reach, the angler fish creates its own light to hunt by. The angler's dorsal fin has evolved into a long and elaborate fishing rod with a luminous end. This lamp, illuminated by tiny

bacteria, dangles just above the angler fish's mouth. In the deep, dark ocean, this lure attracts prey so efficiently that angler fish have no need for a streamlined body shape. Using changes in water pressure, the fish detects the approach of its prey and traps it within heavy jaws filled with long curved teeth.

LIGHT FOR THE BIRDS

Birds use light in different ways. While some use daylight merely to see by, some also use it as a signal for changes in behavior. Others make use of low light levels for night hunting.

Night-time predators like the eastern screech owl depend on their huge eyes for successful hunting. The owl's large eyes gather any light that is available and have evolved to see especially well in very dim conditions.

Most migrating birds use changes in day length as the signal for when to start their long journeys. As the days get shorter during the fall in Europe, the alpine swift flies down to southern Africa. Here, spring is just beginning, and the insects that it catches by darting through the air are plentiful. When the African days shorten, the swift returns northward.

Spring is the season of mating and nest building. Birds such as the ruby-throated hummingbird mate when they sense the days getting longer. This means plenty of food when the new chicks are born.

Sound signals

Throughout the living world, animals use sound to carry messages of one sort or another – some even use it to hunt for prey.

When a bird sings or a frog croaks, it is announcing its presence to prospective mates. The song or call can also carry another message, this time to possible rivals. A male robin's song tells other robins that it has claimed a patch of ground, or a "territory." If a rival male robin ventures into the territory, a fight quickly follows.

Animals vary greatly in their sensitivity to a sound's volume, and also to its pitch. If you have average hearing, the highest pitch you will be able to hear is one that packs in about 20,000 pressure pulses per second (hertz). At the other end of the scale, the deepest pitch that you will be able to make out is a low rumble that consists of about 20 hertz. By contrast, a dog can hear sounds as high as 40,000 hertz, which is why it can respond to a dog-whistle that humans cannot hear.

But even a dog's hearing is far surpassed by animals that use sound to "see" the world around them. When a dolphin or an insect-eating bat hunts for food, it uses pulses of sound to pinpoint

its prey. This system is known as echolocation because the animal homes in on its prey by detecting echoes that bounce back from the prey.

A bat's echolocation, which uses sounds at about 100,000 hertz, works with amazing precision. Bats can fly through fine plastic netting in total darkness even when the gaps are so narrow that they have to pull in their wings. Because sound travels well in water, dolphins and other members of the whale family use an even higher pitch to find their food – up to 250,000 hertz.

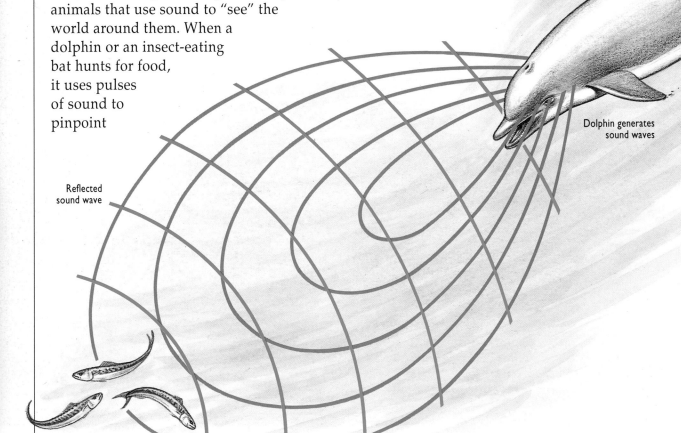

Reflected sound wave

Dolphin generates sound waves

A dolphin uses sound in two distinct ways – to communicate with other dolphins and to hunt. When dolphins make their communication calls, we can hear them as squeaks and clicks. But the sounds that a dolphin uses to track its prey are too high-pitched for the human ear to detect.

Both dolphins and bats use high-pitched sound because it does not scatter easily as it travels outward and will bounce back. High-pitched sound also gives dolphins and bats a detailed "picture" of their prey, because the higher the pitch, the shorter the wavelength. And the shorter the wavelength, the smaller the detail the sound detects.

*Dolphins focus sound into a tight directional beam that can pinpoint prey at some distance. The dolphin has a large brain which is used to process the returning sound signals to give an accurate "picture" of the scene around them, including any prey fish that may be about (**left**).*

ATTRACTIVE SOUNDS

Like many animals, katydids, or large green grasshoppers (**left**), communicate with sound. Katydids are, in fact, among the many insects that use sound to attract their mates. A katydid's forewings have jagged "combs" on their inside surfaces, and these make a sound when the wings are repeatedly scraped together. A katydid detects calls from other members of its species by using "ears," not on its head but on its legs. These grasshoppers are called katydids because the word resembles the noise that they make.

Other insects that attract mates with sound are mosquitoes. When a female mosquito flies, her wings produce sound pressure pulses that we hear as a whine. Male mosquitoes detect the pressure pulses with their antennae. They then track the sound and home in on the source just as a plane homes in on a radio beacon.

On the scent

Smells provide vital early warning signals and help creatures find mates and food, and tell individual animals apart.

If something accidentally catches fire, smell is often the first of your senses to warn you of danger. If you eat something that has begun to turn rotten, your sense of taste prompts you to spit it out at once. Together, these two chemical detection systems can help you to differentiate between situations that are safe and those that are risky or even dangerous.

Humans can distinguish many different kinds of odor, but compared to most other animals, our sense of smell is limited. We can smell things that produce lots of airborne chemicals – such as smoke, freshly mown grass, or hot food – but more dilute forms of scent, such as traces of a drug, escape our notice.

For mammals, reptiles, fish, and insects, smell is often a key way of investigating the world. A dog can detect some scents when they are nearly a million times too faint for us to notice. By picking up different scents, it can tell not only what is close by, but also what is farther away upwind. Smell also tells the dog what has been in the area, and how long ago it was there.

Smell is used not only for detection, but also for identification. Sheep and honeybees are just two of the many species of animal that use personal scent as an identity badge. When a lamb is born, it has a distinctive scent that its mother recognizes, as does an individual honeybee when it is raised in a hive. If a lamb tries to suckle from another ewe, or if a bee tries to enter the wrong hive, the results can be disastrous. Wearing its "foreign" identity badge, the lamb is chased away and the bee is often killed.

For animals that live on dry land, taste and smell are two distinct senses. For aquatic animals they are one and the same, because all the chemicals around them are carried by water. As on land, many animals use these chemicals as signposts. When a migrating salmon leaves the sea and swims upriver to breed, it homes in on the exact chemical "flavor" of the stream in which it hatched. Scientists have found that although salmon may travel vast distances and have to choose between innumerable rivers and streams, they hardly ever make a mistake.

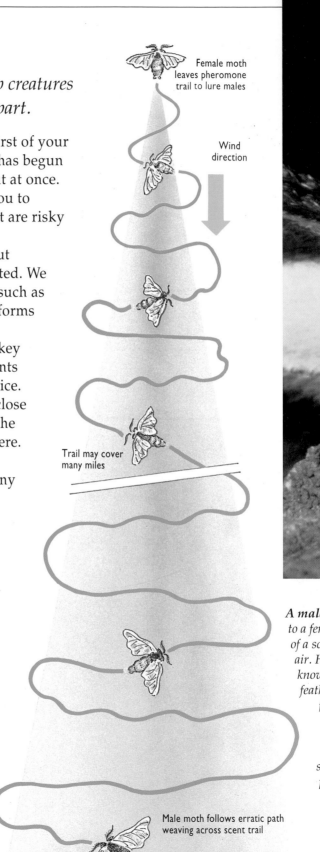

Female moth leaves pheromone trail to lure males

Wind direction

Trail may cover many miles

Male moth follows erratic path weaving across scent trail

A male atlas moth finds his way to a female by sensing tiny amounts of a scent that she releases into the air. He detects the substance, known as a pheromone, with his feathery antennae. When a male first picks up the pheromone, the female may be several miles away. The male tracks down the female by setting off upwind in a zigzag pattern, turning every time the scent trail fades away. By repeating this process, he eventually finds the female.

FOOLING THE FLIES

Many flowers use sweet-smelling chemicals to attract bees which visit the flowers and carry away their pollen for cross-fertilization. But some flowers rely on flies – not normally enticed by flowers – to spread their pollen. The biggest flower in the world, the parasitic *Rafflesia* (**left**) from Southeast Asia, which grows to 3 feet (1 m) across, attracts flies by smelling just like rotting meat – an enticing smell to flies who want to lay their eggs on decaying animal matter. Other fly-attracting flowers have "hairy" petals that feel like the fur of a dead animal.

Strange senses

Birds with a built-in compass and fish that hunt with electricity are using senses outside the usual range.

For hundreds of years, people have navigated using the magnetic compass. The needle of a compass always aligns itself with the Earth's magnetic field, so a fairly straight course can be followed by keeping the needle pointing in the same direction. But some creatures, including migratory butterflies and birds, do not need a compass. For instance, if the weather is cloudy, and pigeons cannot navigate using their normal method of checking the position of the Sun or stars, they have a backup magnetic system. Their ability to sense magnetism helps them find their way as they fly in Earth's magnetic field.

Some animals "feel" their way by sensing a different type of field – one that they set up themselves. In the murky waters of the River Amazon, the knife-fish creates an electric field around itself by generating and emitting low-voltage pulses of electricity. The fish senses anything nearby by detecting the distortions the object makes in that electric field.

The electric eel has taken the use of electricity a step further. It uses a weak low-voltage electric field to detect its food, but then releases an electric charge so powerful that it kills or stuns its prey. The charge is produced in the eel's body in banks of modified muscles called electroplates. These are stacked together like thousands of batteries arranged end to end. Each "battery" generates a small charge at a voltage of about 0.1 volt. When the fish has found its prey, it activates all the cells at once, and together they generate a deadly 500 volts.

An electricity-sensitive mammal is the Australian duck-billed platypus. It does not produce its own electric field, but has sensors that pick up the charge given off by its favorite food, freshwater shrimp.

Insects can also sense electric fields. In a thundercloud, huge electrical charges build up, creating a strong electric field. Bees can sense this, and they become agitated and return quickly to the hive to avoid the storm. Hamsters also sense storms coming and often move their nests for protection. Many plants predict the weather by sensing tiny changes in light, and in the temperature and water content of the air. Scarlet pimpernel, a particularly sensitive plant, closes its flowers tightly just before rain, while clover and wood sorrel fold their leaves.

On migration flights, geese (above) cannot afford to waste any energy taking the wrong route during the long, hard journey. Every winter, barnacle geese migrate, from their summer breeding grounds in the Arctic, south to the salt marshes and estuaries of northwest Europe.

They navigate mainly by vision – using either landmarks, such as coastlines, or by following a bearing, which they set by looking at the position of the Sun or stars. But in cloudy or foggy weather, geese can resort to a backup sense and fly using the direction and strength of the Earth's magnetic field to guide them.

A plant with a sense of touch is Mimosa pudica, *known as the sensitive plant (right). It is famous for the way its leaves suddenly droop if touched. It works by making water rush out of cells at the base of each featherlike leaflet so that it swings downward. Movement in one leaflet triggers the same response in its neighbors, until the whole leaf collapses and looks wilted.*

One theory is that it is a defense against browsing animals. When the animal touches the plant, the leaves droop, making the plant look less appetizing. The animal, in theory, then wanders off, leaving the plant alone.

Fish swimming in a school, like these hatchet fish
(*above*), all seem to change direction simultaneously, as
if responding to a secret signal. This movement, which
confuses predators, happens because fish detect pressure
changes in the water. The detection system, called the
lateral line, is found along each side of the fish's body.
Along the line are clusters of tiny hairs inside cups filled
with a jellylike substance.

   If one of these hatchet fish becomes alarmed and turns
sharply, it causes a pressure wave in the water around it.
This wave pressure deforms the "jelly" in the lateral line
of nearby fish. This moves the hairs which trigger nerves,
and a signal is sent to the brain telling the fish to turn.

   Besides detecting movement in other fish, the lateral
line helps fish find food and build up a picture of the
surroundings. One fish, Astyanax fasciatus, *blind*
because it lives in pitch-black caves, finds food, shelter,
and a mate using its lateral-line detection system.

The generation game

*Producing offspring is the most important function of any species –
otherwise, it would die out. So how do creatures go about it?*

Nothing lasts forever in the living world. As a plant or animal grows older, its life-support systems eventually break down and stop working, and the organism dies. This constant loss of living things is made up for by reproduction. Living things reproduce in two quite different ways – sexually or asexually. A patch of rough grassland in summer may show both systems in action. In the first system, known as sexual reproduction, each grass plant grows flowers which shed pollen grains into the air. When a pollen grain lands on another flower, it unites with a cell inside the flower to produce a seed. The ripe seeds are scattered by the wind and eventually germinate to produce new plants.

The second system, asexual reproduction, is much simpler. In grasses, it involves neither flowers nor seeds, but special stems that creep just below the ground to send up new plants. Asexual reproduction is less complex than sexual reproduction because it involves just one parent. Humans can only reproduce sexually, but many simple animals can reproduce both sexually and asexually. A female aphid, such as a greenfly, for instance, can produce dozens of young without mating with a male.

So why do living things bother with sex? The answer is that asexual reproduction only creates carbon copies (clones) of the parent. This is fine if conditions are good, since all offspring have an equal chance of doing well. But if life gets harder – if, for example, disease breaks out – the result can be disastrous. Since the offspring are genetically identical, if the disease is fatal to one it is likely to be fatal to them all. Sexual reproduction is an advantage because the genes of the two parents mix to give each offspring a unique set of genetic instructions and every offspring is different. If conditions become difficult, the chances are that at least some will survive.

The sexes play different roles in the reproduction and day-to-day life of lions. The females raise the young and do most of the hunting, while the males help themselves to a generous share of the food. This unequal pooling of resources is a common feature of sexual reproduction. In some species of animal – particularly birds – the male plays no part in family life at all, and the female brings up the young unaided.

But there are many systems of family life in which males do play a major part in rearing the young. The emperor penguin, for instance, spends a long period during the Antarctic winter carefully incubating the egg between its feet while the female is elsewhere.

Sea anemones can reproduce both with and without sex, and adult anemones can be male, female, or both. These young anemones (left), clustered around an adult, have developed by sexual reproduction and are growing up under the protection of the parent's tentacles. Anemones reproduce asexually either by splitting into two halves or by detaching small pieces of themselves, which then develop into new animals.

All sexually produced offspring
develop from a single fertilized cell,
or egg. In the green mamba
(***above***), *the egg cell is encased in a*
leathery shell, which protects it and
prevents it from drying out.

Although mammals give birth to
live young, they too have egg cells.
But in mammals the egg cells are
usually tiny and, in most cases,
develop inside the mother's body.
Just three mammalian species – the
duck-billed platypus and the two
types of spiny anteater – lay eggs.

Blueprint for life

The genetic information contained in each living cell is like an encyclopedia of coded instructions – an encyclopedia with the astonishing ability to copy itself.

Like all living things that reproduce sexually, a human being starts life as a single cell. In that cell is a store of chemical information, tiny in size but vast in scope. As the cell divides, possibly to give rise to billions of new cells, exact copies of the information are passed on.

A cell's chemical information shapes its development and regulates every single detail of the way it works. The information is carried in chemical code in giant molecules of a substance called deoxyribonucleic acid, or DNA. A single DNA molecule is like an immensely long ladder that has been twisted to form a double spiral "staircase" (or helix). Each rung of the ladder is formed by two substances, called bases, that lock together. There are four kinds of bases, and together they create four different kinds of rungs. The exact sequence of these rungs, which in humans extends to about 5 billion base pairs, makes up a cell's chemical information.

Genes are individual instructions in the code – a particular series of base pairs that may run from a few hundreds to many thousands. Each gene tells the cell how to make a single protein. Proteins are a wide and diverse collection of chemical substances found in all living things, and they carry out the instructions in the genes.

In most cells, genetic data is carried on several separate DNA molecules housed in the cell's nucleus. Usually they cannot be seen, but when the cell is about to divide, DNA molecules tighten up, or "condense," becoming visible under a microscope. The structures formed are known as chromosomes. Every living thing has a set number of chromosomes in its cells, and all cells, except those involved in reproduction, carry two copies of each. A human, for example, has 46 (23 pairs).

Most genes made of DNA are found within a cell's nucleus. The information needed to control the processes of life is carried in sequences of the four base chemicals cytosine (blue), guanine (orange), adenine (green), and thymine (red), which form the rungs of the DNA molecule. In DNA, each base links only with one other base – cytosine always pairs up with guanine, and adenine always with thymine.

A set of three bases (a codon) specifies a particular amino acid, or sub-unit of a protein molecule. Before instructions in codons can be used to make proteins, an RNA copy has to be assembled.

Cytosine

Thymine

Guanine

Adenine

DNA molecule duplicates itself ready for cell division

*Before a cell divides, its DNA replicates so a full set can be given to each new cell. Enzymes unwind a DNA molecule at many points along its length, separating the two strands. Other enzymes then bring together the bases to build new strands (**left**) to partner each of the originals. The result is two new DNA molecules. The base pairs that link the old and new strands are in the same sequence as in the original DNA molecule, so the cell's genetic information is copied perfectly.*

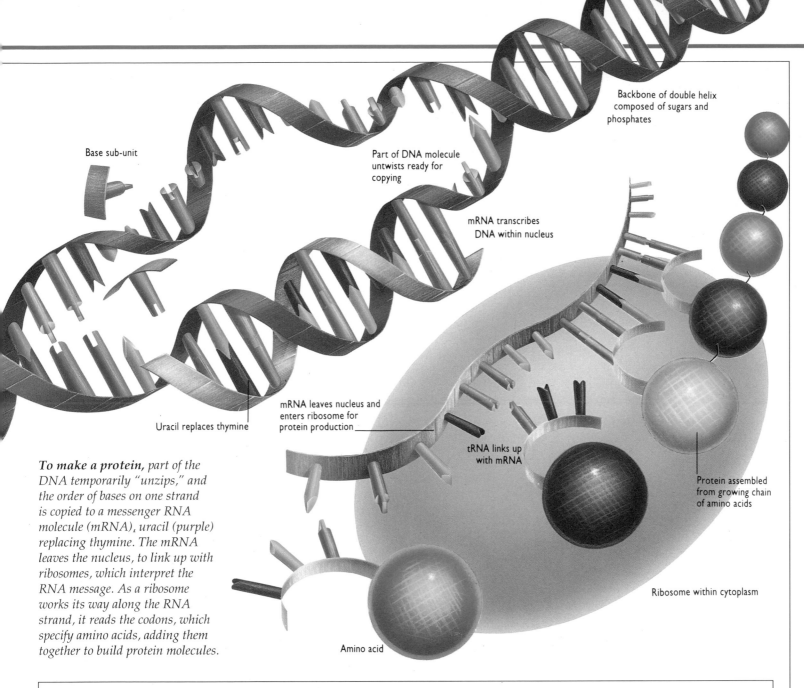

Base sub-unit

Part of DNA molecule untwists ready for copying

Backbone of double helix composed of sugars and phosphates

mRNA transcribes DNA within nucleus

Uracil replaces thymine

mRNA leaves nucleus and enters ribosome for protein production

tRNA links up with mRNA

Protein assembled from growing chain of amino acids

Ribosome within cytoplasm

Amino acid

To make a protein, part of the DNA temporarily "unzips," and the order of bases on one strand is copied to a messenger RNA molecule (mRNA), uracil (purple) replacing thymine. The mRNA leaves the nucleus, to link up with ribosomes, which interpret the RNA message. As a ribosome works its way along the RNA strand, it reads the codons, which specify amino acids, adding them together to build protein molecules.

See also

LIFE
► King carbon 150/151

► The generation game 168/169

► Improving on nature? 172/173

► The microscopic factory 174/175

► The living cell 176/177

► Body building 178/179

SPACE
► Origins of life 16/17

► The struggle for survival 18/19

ATOMS AND MATTER
► Carbon's vital links 130/131

KEEPING IT IN THE FAMILY

Families often show distinctive characteristics that have been handed on from parents to their children. Some characteristics, such as blood groups, are brought about by a single gene, but most result from the combined effects of several. Pure blue or brown eyes, for example, are brought about by alternative forms of one gene. But if "modifier" genes are also present, the blue or brown color is altered to produce one of many different hues.

Improving on nature?

By manipulating genetic information – either indirectly by selection or directly by genetic engineering – creatures can be dramatically changed.

An ordinary garden seed catalogue contains many plant varieties that have never existed in the wild. These varieties are created by artificial selection – a process in which desirable characteristics are picked out and emphasized through controlled breeding. Artificial selection has given us not only showy flowers and high-yield crops, but also a huge variety of animal breeds, from "toy" dogs to beef cattle. But as scientists have come to understand more and more about the mechanisms of inheritance, entirely new ways of changing organisms have been developed. These are collectively known as "genetic engineering," because they rely not on breeding, but on the transfer of individual genes from one organism to another.

In artificial selection, a useful characteristic is passed on as one part of a plant or animal's complete genetic package. Genetic engineering works much more precisely than this. In a laboratory, a "gene map" is drawn up, showing the exact location of the DNA that codes for a particular characteristic. A copy of this DNA sequence can then be chemically removed from a "parent" cell and passed on to a cell that does not already have it.

The extraordinary feature of genetic engineering is that it does not work only with members of

A cultivated rose (right) compared with a wild species vividly shows how living things can be transformed by artificial selection and modern techniques of hybridization. For centuries, generations of rose-breeders have picked out desirable characteristics, such as large flowers, "double" petals, and disease resistance. As a result, there are now thousands of varieties of cultivated rose, each with a different combination of characteristics.

The Hybrid Tea Rose 'Blessings' has delicate pink fragrant blooms. It is the result of breeding together several species of wild rose.

Bacterium

Plasmid

Bacterial chromosome

Human cell

Plasmid cut open using an enzyme

Cell nucleus

Chromosome

The same enzyme isolates and extracts insulin-producing gene

DNA coiled into chromosome

Gene slotted into plasmid

Plasmid inserted back into bacterium

the same species, or even ones that are closely related. Instead, it allows sections of DNA to be transferred between widely different organisms. For example, the DNA that instructs a human cell to make insulin can be transferred into a bacterium. Bacteria divide very quickly, so the insulin-making characteristic is soon passed on to a "clone" consisting of millions of identical offspring. Every cell in the clone is a microscopic insulin factory – an organism that can turn simple chemical food into a valuable, life-saving drug. Bacterial insulin came on stream in the early 1980s and is just one of many extremely useful substances that are now produced by engineered bacteria.

So far, the uses of gene transfer, or recombinant DNA technology, are limited, but many biologists think its effects will eventually be far-reaching. The capability of creating radically new organisms by moving genes between species already exists. However, progress has to be tempered by caution. Genetic engineers have the potential to create dangerous organisms as well as beneficial ones, so research in this area is carefully regulated.

BENEFICIAL BACTERIA

To make bacteria manufacture insulin, an essential hormone that controls the body's use of sugar, the human gene for making insulin is inserted into a structure in the bacteria called a plasmid. The plasmid is cut apart using restriction enzymes and the human genetic information inserted. When the bacteria is cultivated, it then makes "human" insulin.

Modified bacterium multiplies rapidly in fermenter

Human insulin produced

*The "geep" is a genetically engineered chimera – an animal that is made partly of goat cells and partly of sheep cells (**above**). Animals like geeps are made by fusing together dissimilar cells when the creature is an embryo of just a few cells. In a chimera the genetic information in each type of cell belongs to just one species, unlike in a hybrid where the genetic information in each cell comes from two different species.*

The microscopic factory

A cell is the smallest unit able to carry out all life's basic functions. Many single-celled organisms exist, but all large creatures contain millions or billions of cells.

Next time you cut up an onion, try peeling away the thin film from the inside of one of the layers and holding it up to a bright light. If you look closely, you may be able to see faint streaks packed tightly together. Each of these tiny streaks is a strip of cells – the building blocks of all living things.

Compared to most cells, those in an onion's scales are large. An "average" plant cell is about ⅟₅₀₀ inch (0.05 mm) across, so it would take nearly 50 to stretch across the head of a pin. Most animal cells are smaller still, although the single cells that make up eggs are an important and often a gigantic exception. Some living things, such as bacteria, consist of just one cell, but larger organisms, such as trees or humans, contains billions of cells, working closely together.

A cell is like a microscopic chemical factory, carrying out an astounding array of different processes. All these processes are coordinated by the nucleus, which houses the cell's genetic material, DNA. The cell's interior is filled with a watery substance, the cytoplasm, which contains a variety of tiny structures called organelles. These work like a factory's different departments, each specializing in different tasks, so allowing different cells to exist and work together. Surrounding every cell is a special membrane which has pores that let some substances through but block others. This lets the cell maintain its chemical balance.

Protein-lined channel

The cell or plasma membrane *is a thin double layer which encloses the cytoplasm. Sugars, amino acids, and other vital substances pass into the cell through special protein-lined channels.*

Plasma membrane

Lysosome

Nucleus

Microtubule

Nucleolus

Centriole

Cytoplasm

Smooth endoplasmic reticulum

An animal cell *is a small but complex assembly of organelles, the main one being the nucleus which controls and regulates the work of the whole cell. Surrounding the nucleus lies the cytoplasm, a jelly-like fluid which contains water, minerals, and the bulk of the cell's specialized structures, known as organelles. The whole cell is surrounded by a flexible skin, known as the plasma membrane.*

Inside the nucleus *is the command center, the home of the genetic blueprint, DNA. Large pores in the cell nucleus membrane allow copied chemical instructions to pass out into the cell.*

Endoplasmic reticulum *(ER) is a membrane which takes two forms: rough ER makes protein-producing ribosomes; smooth ER makes vesicles – small sacs used to carry material about the cell.*

Centrioles *are precisely arranged bundles of microtubules. They regulate the production of other microtubules and assist in cell division.*

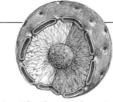

Mitochondria *are the power stations of the cell. These vital structures contain the enzymes that release the energy stored in food.*

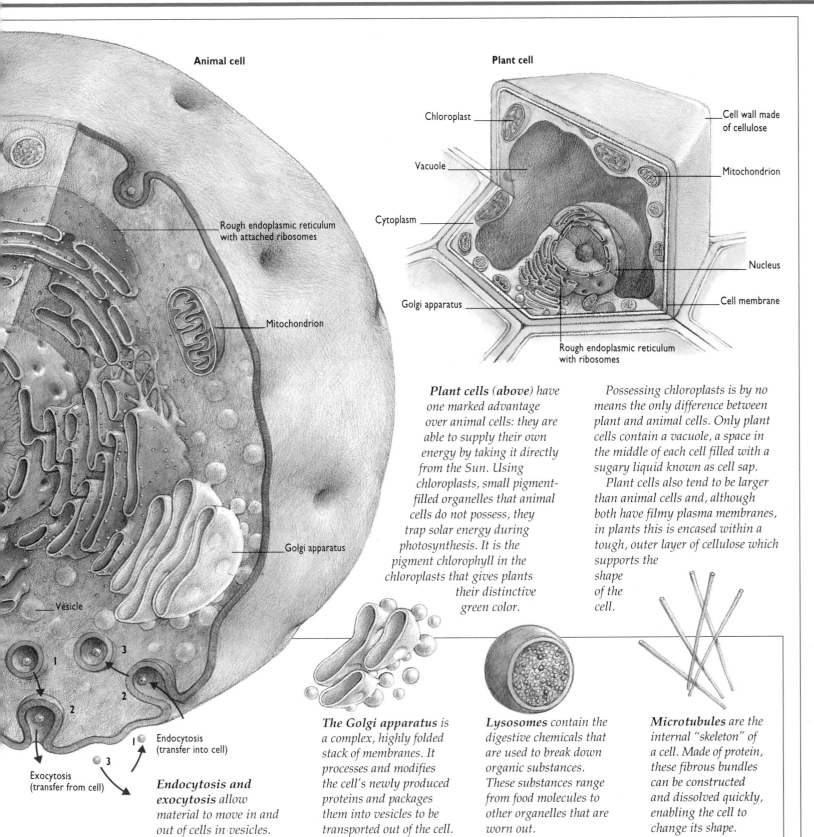

Animal cell

Rough endoplasmic reticulum
with attached ribosomes

Mitochondrion

Golgi apparatus

Vesicle

Endocytosis
(transfer into cell)

Exocytosis
(transfer from cell)

Plant cell

Chloroplast

Vacuole

Cytoplasm

Golgi apparatus

Cell wall made
of cellulose

Mitochondrion

Nucleus

Cell membrane

Rough endoplasmic reticulum
with ribosomes

*Plant cells (above) have
one marked advantage
over animal cells: they are
able to supply their own
energy by taking it directly
from the Sun. Using
chloroplasts, small pigment-
filled organelles that animal
cells do not possess, they
trap solar energy during
photosynthesis. It is the
pigment chlorophyll in the
chloroplasts that gives plants
their distinctive
green color.*

Possessing chloroplasts is by no
means the only difference between
plant and animal cells. Only plant
cells contain a vacuole, a space in
the middle of each cell filled with a
sugary liquid known as cell sap.

Plant cells also tend to be larger
than animal cells and, although
both have filmy plasma membranes,
in plants this is encased within a
tough, outer layer of cellulose which
supports the
shape
of the
cell.

***Endocytosis and
exocytosis*** *allow
material to move in and
out of cells in vesicles.*

The Golgi apparatus *is
a complex, highly folded
stack of membranes. It
processes and modifies
the cell's newly produced
proteins and packages
them into vesicles to be
transported out of the cell.*

Lysosomes *contain the
digestive chemicals that
are used to break down
organic substances.
These substances range
from food molecules to
other organelles that are
worn out.*

Microtubules *are the
internal "skeleton" of
a cell. Made of protein,
these fibrous bundles
can be constructed
and dissolved quickly,
enabling the cell to
change its shape.*

175

The living cell

An adult human body contains about 50,000 billion cells, each of which came from just one fertilized egg.

Every time you rub your hands together, you shed millions of dead cells from your skin, but these cells, like most in your body, are constantly being replaced.

Cells multiply by dividing, which they do in two different ways (**opposite**). The common method, mitosis, allows a cell to create two exact replicas of itself; while the other, meiosis, is only used to create the special cells used in sexual reproduction.

The development of tissues and organs begins after a fertilized cell starts to divide. As division follows division, the total number of cells quickly rockets. The first few divisions occur roughly together, but as the number of cells climbs from hundreds to thousands then millions, some clusters of cells divide more rapidly than others and begin to develop different shapes and characteristics. Some cell clusters form cartilage and bone, while others create muscles, tendons, or body systems, such as the nerves and blood vessels. Eventually, this extraordinary series of divisions forms a recognizable body shape.

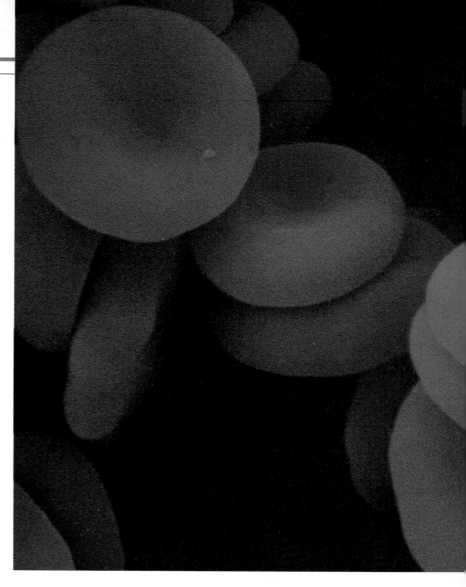

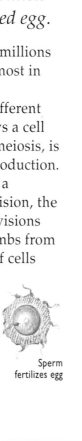

0 days
Sperm fertilizes egg

3 days
Eight cells

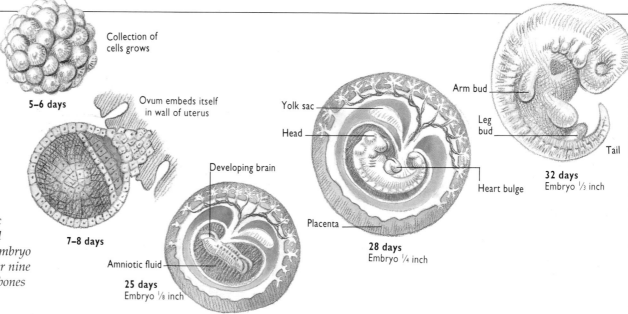

Body development begins at conception – the moment at which a sperm cell from the father reaches and fertilizes the mother's egg.

Within five days, the original fertilized cell has become a hollow ball of cells known as a blastocyst. Hairlike cilia on the lining of the Fallopian tube propel the blastocyst toward the uterus where it implants itself in the lining. Here, the cells in the blastocyst start to divide to form different layers from which a body develops.

By the 25th day, an embryo has begun to form. It has a heart and a tail which is later reabsorbed. The embryo floats in the cavity filled with amniotic fluid, which protects it from jolts. By 10 weeks, cell division and differentiation have transformed the embryo into a fetus with a recognizably human shape. After nine months, the fetus is ready for birth. Gaps between bones in the skull let the head squeeze into the world.

Collection of cells grows

5–6 days

Ovum embeds itself in wall of uterus

7–8 days

Developing brain

Amniotic fluid

25 days
Embryo 1/8 inch

Yolk sac

Head

Placenta

28 days
Embryo 1/4 inch

Arm bud

Leg bud

Heart bulge

Tail

32 days
Embryo 1/3 inch

A single drop of blood contains
about a billion cells, of which the
majority are red blood cells (**left**).
Using a protein called hemoglobin,
these cells carry oxygen from the
lungs to all parts of the body.

Mitosis   **Meiosis**

CELL DIVISION

Mitosis and meiosis
start from a cell (**1**) in
which the chromosomes
in the cell nucleus double
(**2**). In **mitosis**, the cell
splits in half (**3**), so that
each new cell takes one
full set of chromosomes, so
producing two identical daughter cells (**4**).
This process creates the body's tissues after
conception and maintains them during life.

During **meiosis**, the chromosomes double
(**2**) and then form pairs (**3**). At this stage there
is an exchange of some of the genes (**4**). The
cells then split to form two complete, but
slightly different, cells (**5**). These cells rest,
then divide again (**6**), so each new cell has
just half the genetic material needed for
life (**7**). These half cells are the egg and
sperm of sexual reproduction.

See also

LIFE
► The
generation
game
168/169

► Blueprint
for life
170/171

► The
microscopic
factory
174/175

► Body building
178/179

SPACE
► Origins of life
16/17

**BRAINS AND
COMPUTERS**
► Sending
messages
190/191

► Keeping
in touch
200/201

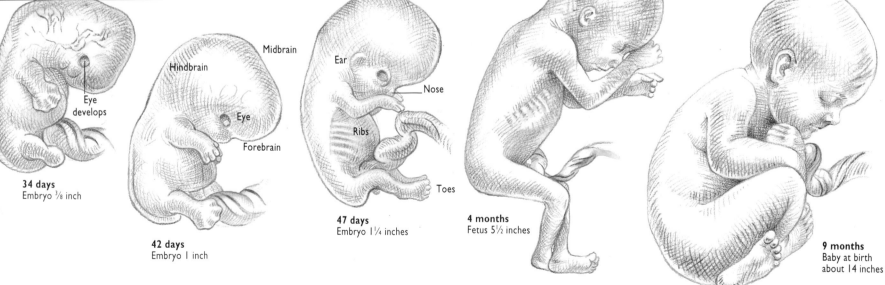

34 days
Embryo ⅜ inch

Eye
develops

42 days
Embryo 1 inch

Hindbrain

Midbrain

Eye

Forebrain

47 days
Embryo 1¼ inches

Ear

Nose

Ribs

Toes

4 months
Fetus 5½ inches

9 months
Baby at birth
about 14 inches

Body building

Some organisms change little in appearance as they increase in size, but others change remarkably as they grow.

When a butterfly slowly hauls itself out of its chrysalis, it completes one of the most astonishing transformations in the living world. The animal that makes the chrysalis is a caterpillar – a slow-moving, tube-shaped, wingless eating machine with small but powerful jaws. The animal that emerges is quite different – a fragile and brightly colored flying insect that drinks nectar delicately through a tubelike tongue.

In human development, this total change in body shape is inconceivable. We do change shape as we grow, but gradually and in a limited way. An adult human, unlike a child, has working reproductive organs and just one set of teeth, but the main differences between adults and children lie in body proportions, rather than in basic body plan.

Animals whose body patterns change dramatically as they grow are said to undergo metamorphosis. These animals include not only butterflies, but many other invertebrates (animals without backbones), as well as amphibians such as frogs and toads. By having more than one body shape, these animals can make use of more than one food source and can also cover a greater distance more easily, thus giving the species a greater chance of survival. Adult barnacles, for example, are protected by hard chalky armor and are glued to a rock. They cannot move, but their young, called larvae, drift in the sea and are often carried far away from their parents. In the same way, a butterfly can roam far away from the place in which it grew up as a caterpillar.

Animals usually stop growing once they have reached maturity, but many plants continue to grow whenever conditions are favorable. Like animals, they also change shape as they grow, but their form is very much influenced by their surroundings. For instance, on a tree planted in a windy place, the buds facing into the wind often dry out and die, so the tree grows mainly in the opposite direction. The same tree in a sheltered spot looks quite different. Light levels can also have an important effect on a plant's shape. If a potato is forced to sprout in dim conditions, it develops long stems as it seeks out the light. In bright light, the same potato forms a much more compact plant.

Newly emerged from its chrysalis, a butterfly looks vastly different from the caterpillar it grew from, yet the two forms are both the same creature. A caterpillar enters metamorphosis when its body stops making a substance called juvenile hormone. After the caterpillar has formed a chrysalis, most of its cells are broken down to form an organic soup. At the same time, pockets of other cells begin to divide, using the chemicals released by tissue destruction. These dividing cells eventually form the organs of the butterfly's body.

WHAT A WAY TO GROW

Organisms change in different ways as they grow. In humans a newborn baby has a large head in proportion to its body. By 1 year, the limbs are larger and are sturdy enough for crawling. By 6 years old, only about one-fifth of the height is made up by the head, and other parts of the body – especially limbs – grow fast. By 13 or 14, sex hormones trigger a period of rapid growth when boys grow facial hair and girls' hips widen. By the age of 20, the body's growth is usually complete.

The tortoise is a creature whose shape stays the same although every one of its components enlarge. Each plate in its shell increases in size as the creature grows, but the proportions stay much the same during its lifetime.

As a snail grows, it adds new material to its shell. The shell grows only along its lip; more turns are added, but existing turns are not enlarged, and the snail moves into the new space created. Alternating periods of fast and slow growth create bands in the shell, like the seasonal growth rings seen in a tree.

Completely contained by its hard outer shell, called an exoskeleton, a crab grows in dramatic spurts when it sheds the shell. For a short time after shedding the old shell, the new shell is soft, enabling the crab to expand rapidly in size before the new shell hardens.

In ideal conditions, bamboo can gain height at a rate of nearly 3 feet (1 m) in 24 hours, making it the fastest growing of all land plants. This growth is produced partly by cell division and partly by cell enlargement.

The environment in which bamboo grows provides it with the nutrients and energy it needs to make such rapid progress from a tiny seedling to a fully developed plant in a short time. Bamboo needs to grow in regions where there is plenty of energy from sunshine and plenty of water available in the soil.

Most bamboos are thus found in tropical or semitropical regions of the world. The bamboo here, growing in the lush Indonesian island of Java, is called Dendrocalamus giganteus.

179

The living alliance

Living together can be like love or war – helpful and supportive, or destructive and deadly.

In nature, the need to survive and succeed is as unrelenting as in the world of business. Just like people in business, living things seize any opportunities that improve their chances of success. Sometimes this success comes not through battling against all comers single-handedly but through forming partnerships.

Animals and plants form partnerships either with members of their own species or with members of another. In some mixed-species partnerships, both partners gain something useful from each other, and the arrangement is mutually beneficial; this type of relationship is called symbiotic, or mutualistic. Many flowering plants, for example, supply honeybees with food, in the form of nectar and pollen. And as bees feed, they help the plants by carrying pollen from one flower to another to fertilize the plants.

Partnerships are the result of evolution, and as generations succeed each other, the balance between partners can change. Frequently, one species will start to take advantage of the other, and what started as an equal partnership can become completely one-sided. This situation is known as parasitism.

A parasite is an organism that makes its living at the expense of another. Parasites take many forms and are found in almost every major branch of life. They include parasitic plants and fungi, single-celled organisms such as the one that causes malaria in humans, and a vast range of parasitic invertebrates (animals without backbones). One parasitic vertebrate, the lamprey, feeds on other fish by clamping its circular mouth on to its victims and then rasping at their flesh.

Loathsome though parasites may seem, they are often superbly adapted for their way of life. Tapeworms, for instance, drain their hosts of resources in a most efficient way. Through evolution, tapeworms, which live in their hosts' intestines, have lost all the organs they do not need, so they have no digestive system. Instead, a tapeworm's body absorbs already digested food directly and turns it into eggs. A single tapeworm can release over a million eggs a day for months on end, maximizing the chances that at least some will find their way to new hosts.

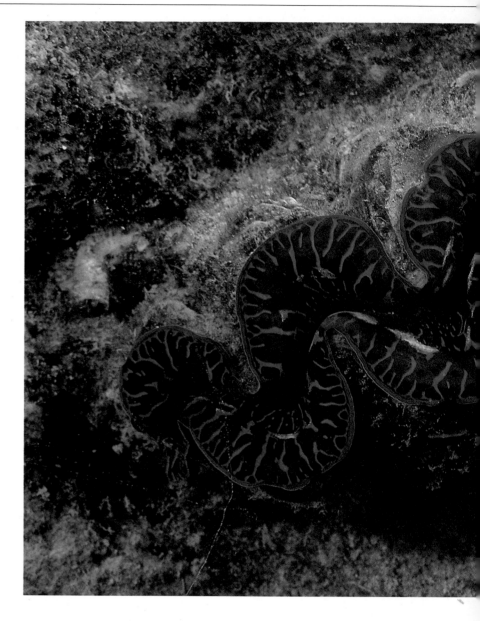

One partnership in which species gain from each other has the giant clam, Tridacna gigas, *as its senior partner. Giant clams (***above***) are found in warm, shallow water in the tropics. They feed partly by filtering small particles of edible organic matter from sea water, and partly by exploiting the tiny plants that live inside their soft tissues. The plants, called zooxanthellae, make food by harnessing the energy in sunlight using photosynthesis. Some of the food they make is used* by the giant clam. The plants gain from the alliance because in return the clam protects them.

As in all partnerships, there is no deliberate give-and-take: each organism simply makes the most of the other. In the case of the clam and the plants, both have enhanced chances of survival, thanks to their mutually beneficial relationship. If they survive to breed successfully, then the partnership between the two species is more likely to recur in following generations.

See also

LIFE
▶ Light for life
154/155

▶ The
generation
game
168/169

▶ Working
together
182/183

▶ Networks
for life
184/185

SPACE
▶ The struggle
for survival
18/19

ENERGY
▶ Beyond violet
76/77

**BRAINS AND
COMPUTERS**
▶ Getting the
message
202/203

A parasite that steals its food is the dodder (**right**), a climbing plant that has tiny scalelike leaves. It starts its life in a straightforward manner, by germinating from its seed and scrambling up nearby plants. But if its stems come into contact with a suitable host, they burrow into the host plant and start to rob it of nutrients. Once it has latched onto a free source of food, the dodder's roots wither away, and its spaghetti-like stems grow into a thick mat draped over the host.

Many grazing and foraging animals, such as this warthog (**above**), are plagued by small animals that live on their skin by sucking blood. A warthog cannot remove these ectoparasites, but insect-eating birds, such as the red-billed oxpecker, can. The birds gain from the relationship by searching out and eating the insects. The warthog gains by being freed from the ectoparasites and so allows the birds an easy meal by not chasing them away.

Working together

Creatures ranging from ants to humans live in groups cooperating to provide food and shelter for all.

The air around a mature, basketball-sized wasp nest is often thick with wasps purposefully flying to and fro in search of food. This kind of nest is a complete society of closely related animals living and working together. Among insects, social living has evolved not only in wasps but also in bees and ants, and in termites. Their societies are often described as "super-organisms" because, in many ways, an entire nest works as a single animal. Only one individual – the queen – lays eggs. The other members of the nest – the workers – carry out a variety of duties, including collecting food, repairing the nest, keeping it warm, and bringing up the young.

Many other animals also live in groups, but do not divide up the work of staying alive. Antelopes, for example, have a better chance of escaping predators by living in a herd, but there is no real division of labor. Each adult antelope finds its own food, and all the adult females can produce young.

There is just one mammal whose strange lifestyle does seem to have something in common with social insects – the naked mole rat, of eastern Africa. Naked mole rats live in groups of up to 100 animals. In each group, only one of the females breeds, and this "queen" is supported by workers that find food.

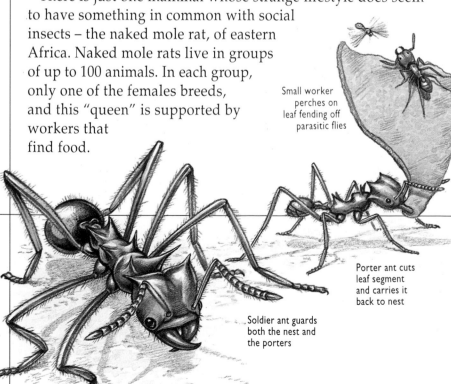

Small worker perches on leaf fending off parasitic flies

Porter ant cuts leaf segment and carries it back to nest

Soldier ant guards both the nest and the porters

INSECT SOCIETY

Seemingly simple creatures live in complex groups. Leafcutter ants live in societies where not only are the different roles clearly defined, but the shapes of the ants performing each role are different. Large female workers cut the leaves while guarded by even larger soldier ants. Smaller workers perch on the cut leaves fending off parasitic flies. The cut leaves are stacked below ground, in compost piles that nourish a species of fungus. The ants, which cannot digest leaves, use the fungus as food.

Commuters swarm along a railroad platform, intent on the activities of the day ahead (**left**). Most humans live in extremely complex social systems in which individuals have highly specialized roles such as engineer, accountant, chef, construction worker, doctor, and journalist.

Humans are, in fact, a highly social species, but among scientists opinions vary about how far human behavior and society can be explained in strictly biological terms. In one sense, the people living in a city are members of small breeding groups that vie with each other for resources. But these groups are not entirely independent, because in a community such as a city, every individual depends on the activities of thousands of others. Competition therefore goes hand in hand with cooperation.

Corals (right) are soft-bodied animals that often surround themselves with a stony case. They show a complete range of lifestyles – some live entirely alone, while others form the large, tightly packed colonies that make up coral reefs. Unlike ants, corals do not completely give up their independence when they live together. Each coral animal, or polyp, keeps its ability to reproduce, although it is sometimes difficult to know where one polyp ends and its neighbors begin.

See also

▶ **LIFE**
The generation game
168/169

▶ The living alliance
180/181

▶ Networks for life
184/185

SPACE
▶ The struggle for survival
18/19

BRAINS AND COMPUTERS
▶ Getting the message
202/203

▶ The higher functions
212/213

An ant nest may contain millions of workers, but it has only one egg-laying queen, tended by small ants whose task is simply to attend to the queen's needs so she can lay eggs to continue the colony.

Worker ant prepares leaves

Small worker ant turns leaves into balls of pulp and places them in fungus garden

Gardener ant tends the growing fungus

Queen ant lays eggs

Networks for life

Energy passes in a never ending stream from one living thing to another along food chains. Find out how a top predator is linked to a primary producer.

Like all forms of energy, the energy that keeps living things alive and working can neither be created nor destroyed. However, it can be passed on. When a plant captures energy from sunlight during the process of photosynthesis, it acts as the first link in a food chain that will eventually pass energy through several different living things.

When an animal eats a plant, or an animal eats another animal, the energy in the food moves forward one link in the chain. The plant at one end of the chain is known as the primary producer because it alone brings about the supply of energy from sunlight. The animals that form the other links in the chain are the consumers. Plant-eating animals, or herbivores, are the primary consumers; carnivores form further layers of consumers by feeding on the plant eaters.

Since there are so many species of living things in the world, it might seem likely that food chains would have hundreds or perhaps thousands of links. But food chains are far shorter than this. Most of them connect one plant with perhaps three or four species of

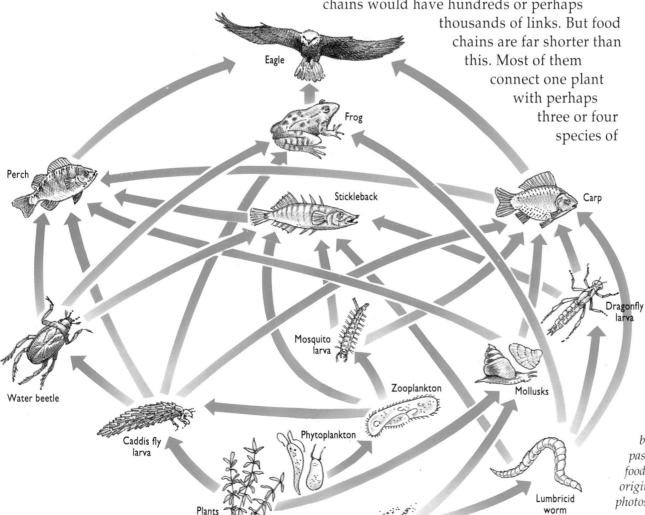

Eagle
Frog
Perch
Stickleback
Carp
Dragonfly larva
Mosquito larva
Zooplankton
Mollusks
Water beetle
Caddis fly larva
Phytoplankton
Lumbricid worm
Plants
Detritus

*Atop an aquatic food web, the fish eagle (**left**) preys on large fish and amphibians such as the frog. There are often only five links in a food chain, but interlinking chains make a complicated web.*

When an animal eats food, about 10 percent of the food's chemical energy becomes chemical energy in its body. The rest – 90 percent – powers body processes and therefore cannot be passed on to another animal. In a five-link food chain, only one ten-thousandth of the original energy stored by plants during photosynthesis is passed to the top predator.

A fish eagle (left) is an animal at the top of several food chains. Although a young fish eagle may fall victim to predators, an adult bird has no natural enemies, making it a top predator. The food chains completed by the eagle bridge the realms of land and water.

Top predators such as fish eagles often provide early warning signs of environmental change. They soon suffer if their food chains are disrupted, and they also tend to accumulate manmade chemicals that have been stored in the bodies of their prey.

Without decay, the world would soon become filled with dead plants and animals, and food chains would come to a halt. Fortunately, organisms such as bacteria and fungi, known as decomposers, are constantly breaking down the leftovers of life and converting them into raw materials that can be used once more. But not all fungi and bacteria wait until their food is dead. The porcelain fungus (below) feeds on living cells in a tree.

See also

LIFE
▶ Order from energy 152/153

▶ Light for life 154/155

▶ The living alliance 180/181

▶ Working together 182/183

SPACE
▶ The struggle for survival 18/19

ENERGY
▶ Harnessing energy 64/65

animal. The reason for this is that only a part of the energy an animal takes in goes into building its body. Much more goes into powering the body and making it work, and this energy cannot be passed up the chain. By the time energy has made its way from a plant to a mouse, then to a bird, then to a fox, so much is lost that there is not enough to support any more links.

Every species has its own unique way of fitting into its habitat – the world immediately around it – and making use of the food available. This is known as its ecological niche. Because the food used by different species overlaps, the inhabitants of an area are connected by food webs that bind them together. While food chains are quite simple, food webs can be extraordinarily complex, and their effects extremely subtle. When humans interfere in a food web – for example, by using chemicals to control a pest – many other species in the web may suffer as a result.

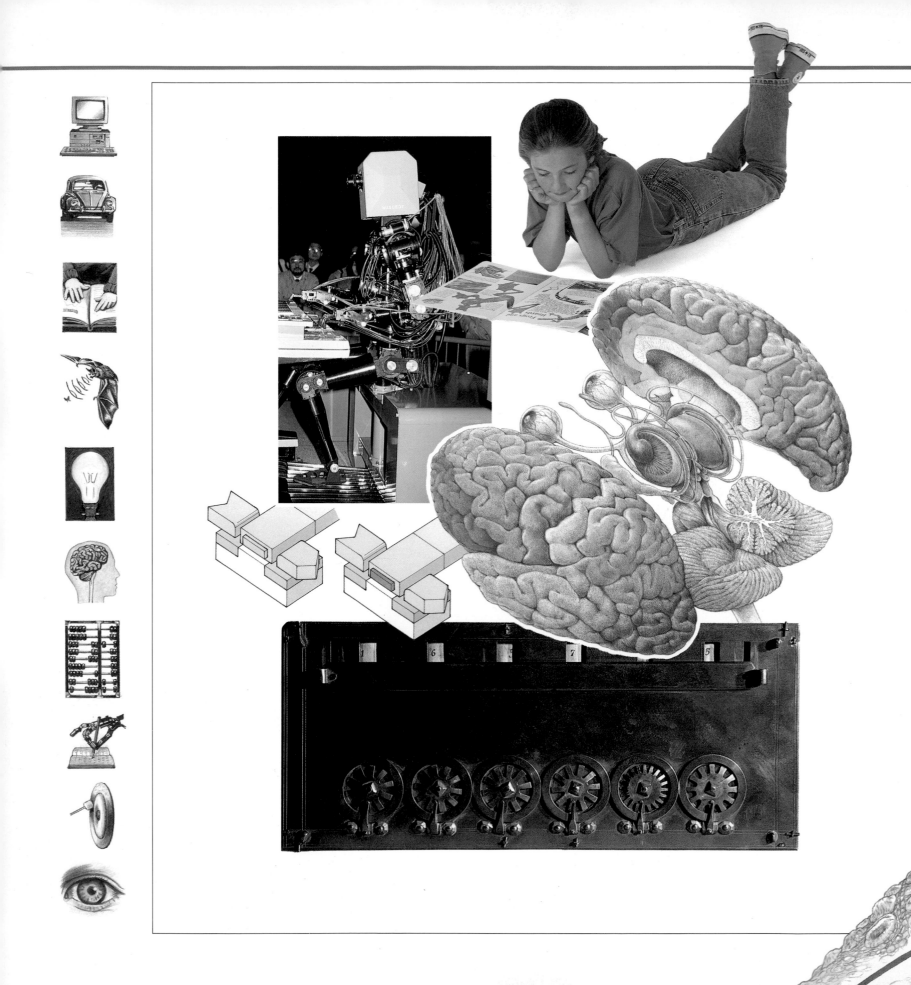

Brains and Computers

Our fabulously intricate and capable brains handle complicated images, words, numbers, and ideas almost without thinking. We have even used the ingenuity of our brains to devise machines and methods that are revealing how the brain itself works. Today's techniques identify the parts of this huge network of nerve cells involved in perception, memory, thinking and speaking, in emotions, and in consciousness itself.

But a startling revolution is going on. At last the human brain is being rivaled by machines that some say will one day ape its every function. The technology of computers has evolved rapidly, and computers and robots, their active counterparts, are invading ever more areas of daily life.

Left clockwise from top: *a robot plays the piano; reading and handling ideas; a look inside the enormously complex human brain; the first mechanical calculator; the silicon heart of the computer.*
This page (top): *the eye, our prime means of learning about the world;* (**left**) *the brain of an alligator.*

Making brains

Internal command and control systems are vital to the lives of complex creatures. The more complex the creature, the more sophisticated the system.

In 1877, Alexander Graham Bell's telephone was a great improvement on shouting messages. Its electrical signals traveled along wires about 900,000 times faster than sound in air (as do phone calls today). They could go as far as a wire carried them, rather than fading over shouting distance.

At once, people clamored for telephones. But connecting each phone to every other would have produced a giant, inefficient web of wires. Engineers devised a system of exchanges, as meeting points for all local telephone wires.

But more than 500 million years before Bell, nature evolved its own version of the telephone wire: the nerve. The advantages of the nerve were the same as the advantages of the telephone wire. Messages could be sent quickly, over

relatively long distances inside an animal's body. And similar to the electrical pulses of a telephone message passed down a wire, a nerve transmits tiny electrical impulses, known as nerve signals.

This communication system gave animals a much improved chance of survival. It was faster than systems that relied on diffusing chemicals or randomly spreading electrical impulses, which soon got lost in a big body. Early many-celled creatures such as jellyfish probably had simple netlike systems of nerves with no central control region.

Time saw a broad evolutionary trend to bigger, more complex animals, with many different muscles and body organs doing specialized jobs. These parts needed

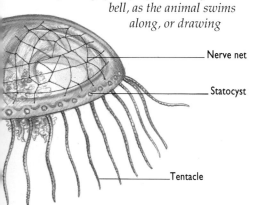

Simple animals like jellyfish have a nerve net, but no central brain. The net coordinates a few uncomplicated reactions and movements, such as controlling the pulsing of the jellyfish's body, or bell, as the animal swims along, or drawing

Nerve net

Statocyst

Tentacle

tentacles, touched possibly by some piece of food, to the mouth.

Other simple parts of the jellyfish's nerve or perception system are statocysts, fluid-filled sacs that are organs of balance.

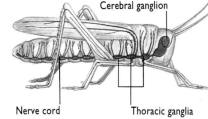

Cerebral ganglion

Nerve cord

Thoracic ganglia

The nerve system of an insect (*above*) consists of three main components. First is a brain in the head, called the cerebral ganglion, closely connected to the eyes, antennae, and other sense organs. The brain is linked to a series of nerve knots called thoracic ganglia, which help coordinate and control the functions in the thorax, or chest region, of the insect.

A large nerve, called the ventral nerve cord, runs along the base of the body, sending branches to the legs, guts, and other parts. The average insect brain has between 10,000 and 100,000 nerve cells.

The brain of a reptilian vertebrate (animal with a backbone) such as an alligator is made up mainly of a brain stem and cerebellum, with a small cerebrum or central coordinating region. The brain stem controls vital processes such as breathing and heartbeat. The cerebellum is a center for muscle coordination.

There are also well-developed regions for important senses. One is the optic lobe, which deals with nerve signals from the creature's eyes, and the other is the olfactory bulb, handling the alligator's sense of smell.

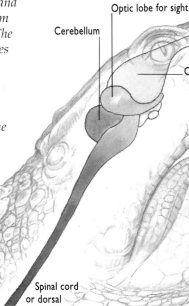

Olfactory lobe for smell

Optic lobe for sight

Cerebellum

Cerebrum

Spinal cord or dorsal nerve tract

coordination. So some nerves became involved in this control function for better internal working and faster, smoother muscle movements.

More senses for monitoring the surroundings became an advantage – especially for land creatures. Some nerves evolved endings on the body's surface that detected touch, chemicals, or other features, providing information about the external world. These were located at the front of the animal – the first part of it to enter new territory.

Along with better senses and coordination, nature evolved its own "telephone exchange": a single, controlling brain, networked via the nerve system into the whole body. Evolving a brain enabled certain animals to get ahead in the struggle for survival.

Most animals today have a three-part nerve system made up of sensory nerves, brain, and motor nerves. Signals from sense organs travel along sensory nerves to the brain, where they are processed and analyzed. Signals are then sent from the brain along motor nerves to muscles, to make possible coordinated actions.

The brain and associated main nerves are the central nervous system. The sensory, motor, and other nerves spreading through the body are the peripheral nervous system.

Median nerve

Ulnar nerve

Compared to the brains of other animals, the human brain (**below**), which contains about 100 billion nerve cells, is dominated by the cerebrum. This contains the cortex, the site of higher mental activities such as reasoning. More "primitive" regions of the brain such as the brain stem, hypothalamus, limbic system, and thalamus are smaller compared to those of other creatures. Reaching down from the brain via the spinal cord is the complicated network of nerves (**right**).

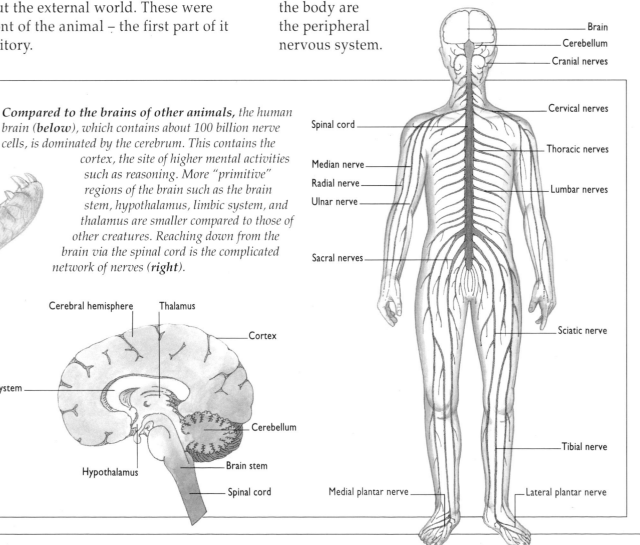

Brain
Cerebellum
Cranial nerves

Cervical nerves

Spinal cord

Thoracic nerves

Median nerve

Radial nerve

Lumbar nerves

Ulnar nerve

Sacral nerves

Sciatic nerve

Cerebral hemisphere | Thalamus

Cortex

Limbic system

Cerebellum

Brain stem

Hypothalamus

Spinal cord

Tibial nerve

Medial plantar nerve

Lateral plantar nerve

Sending messages

The key components of the body's communication system are nerve cells. They transmit internal messages as tiny electrical pulses.

The basic unit of a telephone system is the telephone itself, together with its trailing wire. The mouthpiece changes the sound of a voice into electrical signals and sends them along the wire, into a network which takes them to their destination.

Like the telephone network, the nervous system is built around a basic unit called the nerve cell or neuron. Although neurons are concentrated in the brain, they are found throughout the body, sending, receiving, and interpreting countless messages. The nervous system of a tiny worm has several thousand neurons. Humans have billions.

The "telephone" part of a typical neuron is the cell body. This has many long, thin filaments called dendrites branching from it. The "wire" part is a much longer, thicker extension known as the axon, which has its own end branches. A key feature is that the ends of the dendrites and axons are linked to other neurons by connections known as synapses.

Neurons are specialized to conduct nerve signals, and the synapses pass the signals from one neuron to another. A nerve

STIMULUS AND RESPONSE

As the parent approaches the nest, the chick's eyes detect changing light levels and turn this into coded nerve signals, sent along neurons in the optic (eye) nerve to the brain. The brain parts specialized for dealing with information from the eyes are the optic lobes. Signals showing that a shape has approached trigger a series of intermediate neurons in the cerebrum.

*Nerve messages are carried by axons (**below**), offshoots of neurons swathed by Schwann cells. A synapse (detail, **left**) is where a neuron's axon links to another neuron. A gap of just 0.000008 inch (0.0002 mm) separates them. The nerve message is electrical until it reaches the end of the axon, called the terminal. There it jumps the gap in the synapse as a chemical called a neurotransmitter, which is picked up by receptors in a spine on the dendrite. It continues in electrical form along the dendrite of the next neuron.*

Dendrite
Dendritic spine
Receptor
Synaptic gap
Direction of nerve impulse
Neurotransmitter released

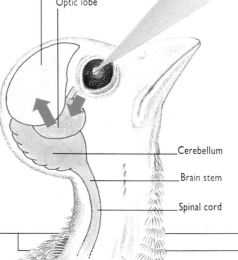

Cerebrum
Optic lobe
Cerebellum
Brain stem
Spinal cord

Nucleus
Nucleolus
Dendrite
Axon
Myelin sheath
Axon hillock
Schwann cell
Myelin sheath
Axon

Baby birds are almost helpless, and they certainly cannot find their own food. But they have instinctive or built-in actions, such as the gaping response – in which they open their beaks wide – when the parent arrives at the nest with food.

When the baby bird realizes that the parent has arrived, it opens its mouth wide. After transmission to the cerebrum, the message is relayed to the cerebellum which specializes in organizing and coordinating muscle actions. The gaping motion is thus controlled by the baby bird's cerebellum, which sends a set of signals along the motor neurons to the muscles in the head, jaws, and neck. These tip back the head and open the beak, so the chick's mouth gapes.

The parent sees the colorful gaping mouth. A similar series of events happens in its eyes, brain, and muscles, triggering the feeding reaction.

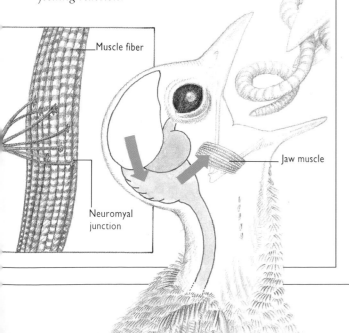

Muscle fiber

Jaw muscle

Neuromyal
junction

signal can be measured with sensitive electrical equipment. It has a strength of 30–80 millivolts (about one-thirtieth of an average battery), lasts less than one-thousandth of a second, and flashes along a neuron at speeds of up to 330 feet per second (100 m/sec).

Nerve signals travel not through the cell's interior, but along its outer skin or membrane. The signals come into the neuron through the dendrites, pass along the cell body, travel along the wirelike axon, and leave at the axon terminals, passing to the dendrites of other neurons. The electrical signals in axons are insulated by sheaths of cells called Schwann cells.

The long, cord-shaped nerves in an animal's body consist mainly of axons bundled together. For example, a nerve in your toe, thin as a thread, has several thousand axons. Despite their thinness, some axons are longer than 20 inches (50 cm). They pass up to the spinal cord in the base of your back, where they connect via synapses to more neurons in the cord.

The numbers of neurons and the synapses between them make the system incredibly complex and adaptable. In a human brain, there are some 100 billion neurons. One neuron may have tens of thousands of dendrites. Each dendrite may have synapses with tens of thousands of other neurons. The overall complexity and the possible pathways for nerve signals are literally mind-boggling.

The brain: head office

Packed snugly into the space between your ears is perhaps the most remarkable device in the universe – the human brain.

Computers are marvels of the modern age. A lap-top computer weighing around 7 pounds (3 kg), with appropriate programs and disks, can store the entire writings of William Shakespeare or several dozen color images.

An object less than half the weight of the lap-top far outstrips these data-storage and handling abilities. Pinkish-gray in color, and with the texture of yogurt, the human brain stores a lifetime of information about sights and sounds, smells and tastes, thoughts and ideas, feelings and emotions.

The human brain's processing power almost defies imagination. For instance, in a helicopter the pilot is swamped by input, or sensory information: the view around, flight displays, engine sounds, headphone messages, motion and balance transmitted through the seat, the feel of the hand and foot controls. Each second, more items of information come in through his senses than pass through a large town's telephone exchange in a week.

Output is no less startling. Millions of nerve signals each second coordinate the 600-odd body muscles as the pilot manipulates the control stick and foot pedals, moves his head and eyes, and talks into the microphone of his radio.

Between input and output is the astonishing activity inside the brain itself. It sifts and analyzes inputs, allocates priorities, considers choices, and decides actions. A limited amount happens in the pilot's awareness; much more is automatic, or practiced and rehearsed.

__Under the wrinkled roof__ of the cortex, the brain is far from being a featureless mass of nerve cells (neurons). It has distinct parts, nerve tracts, lumps, and chambers.

For instance, below the cortex is the curved shape of the limbic system, involved in memory, feelings and emotions, and sexual behavior. The hippocampus helps with transfer, storage, and recall of memory.

To the rear is the cerebellum, whose wrinkled lobes make it look like a mini version of the cerebral hemispheres. The cerebellum organizes and coordinates muscle actions and reflexes, and receives some sensory inputs.

The medulla is like an automatic pilot. It organizes activities such as heartbeat, breathing, and chewing, without you having to think about them. It contains many neurons connecting to the spinal cord below, and thence to the rest of the body.

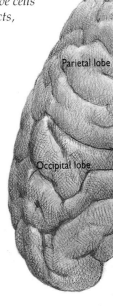

Parietal lobe

Occipital lobe

Cerebellum coordinates movement

THE FEELING BRAIN

About two-thirds of the way back from the front of the head are the twin bands of cortex responsible for analysis and perception of the sense of touch, called the somatosensory cortex. Each side of the brain deals with the senses of the other side of the body. Thus the left side of the brain perceives touch on the right side of the body. This is true for all other perception and control areas: one side of the brain deals with the opposite side of the body.

The areas of the somatosensory cortex that deal with areas of the body are not of equal size. Thus sensitive regions like the fingertips have a much larger area of cortex devoted to them than do less sensitive areas such as the thigh. The figure (**right**) shows the body parts drawn in proportion to the somatosensory areas that represent them.

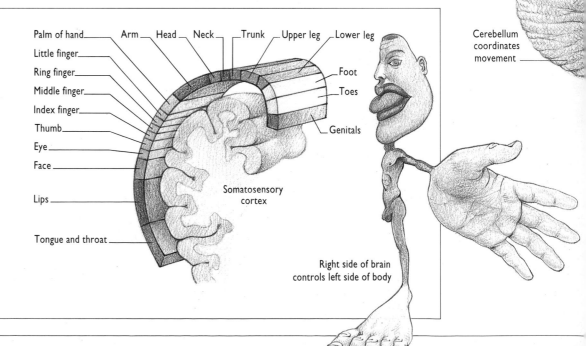

Palm of hand — Arm — Head — Neck — Trunk — Upper leg — Lower leg

Little finger

Ring finger

Middle finger

Index finger

Thumb

Eye

Face

Lips

Tongue and throat

Foot

Toes

Genitals

Somatosensory cortex

Right side of brain controls left side of body

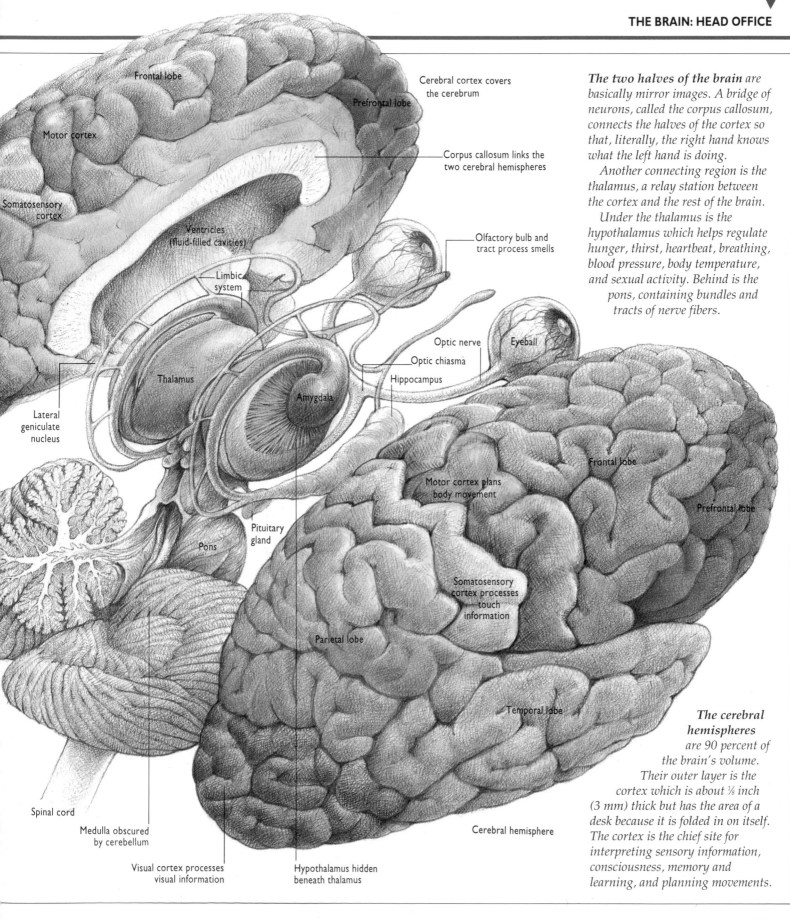

Frontal lobe

Prefrontal lobe

Cerebral cortex covers
the cerebrum

Motor cortex

Corpus callosum links the
two cerebral hemispheres

Somatosensory
cortex

Ventricles
(fluid-filled cavities)

Olfactory bulb and
tract process smells

Limbic
system

Optic nerve

Eyeball

Optic chiasma

Thalamus

Hippocampus

Amygdala

Lateral
geniculate
nucleus

Frontal lobe

Motor cortex plans
body movement

Prefrontal lobe

Pituitary
gland

Pons

Somatosensory
cortex processes
touch
information

Parietal lobe

Temporal lobe

Spinal cord

Medulla obscured
by cerebellum

Cerebral hemisphere

Visual cortex processes
visual information

Hypothalamus hidden
beneath thalamus

The two halves of the brain are
basically mirror images. A bridge of
neurons, called the corpus callosum,
connects the halves of the cortex so
that, literally, the right hand knows
what the left hand is doing.

Another connecting region is the
thalamus, a relay station between
the cortex and the rest of the brain.

Under the thalamus is the
hypothalamus which helps regulate
hunger, thirst, heartbeat, breathing,
blood pressure, body temperature,
and sexual activity. Behind is the
pons, containing bundles and
tracts of nerve fibers.

**The cerebral
hemispheres**
are 90 percent of
the brain's volume.
Their outer layer is the
cortex which is about ⅛ inch
(3 mm) thick but has the area of a
desk because it is folded in on itself.
The cortex is the chief site for
interpreting sensory information,
consciousness, memory and
learning, and planning movements.

See also

**BRAINS AND
COMPUTERS**

▶ Making brains
188/189

▶ Sending
messages
190/191

▶ Window on
the world
194/195

▶ Sound sense
196/197

▶ The chemical
senses
198/199

▶ Keeping
in touch
200/201

▶ The
conscious
mind
204/205

▶ Mapping
the brain
208/209

▶ Total recall
210/211

▶ The higher
functions
212/213

LIFE

▶ Keeping
your balance
156/157

▶ Messengers
of life
158/159

Window on the world

Sight has become our dominant sense – so important that probably two-thirds of the brain's capacity is occupied just with interpreting what we see.

In some ways, the eye is a little like a video camera. At the front is the cornea and lens to focus light and form a tiny picture or image. Lining the back of the eye is the retina, a wafer-thin array of 130 million or so light-sensitive rod and cone cells which detect the image and translate it into a series of nerve impulses, just as the light-sensitive cells in a video camera generate an electrical signal. These nerve impulses are processed by various groups of neurons (nerve cells) and then flashed to the brain along the optic nerve.

Inside the brain, the optic nerves from both eyes meet at the optic chiasma, where neurons carrying signals from the right half of each retina join up, as do neurons carrying signals from the left half of each retina. So each side of the brain receives signals from the corresponding half of each retina.

The signals continue from the optic chiasma along a nerve tract to the next staging post, which is called the six-layered lateral geniculate nucleus. Here, the signals split up and fan out to an area called the visual cortex on the lower rear section of each half of the brain.

Experiments with animals show that different regions of the visual cortex analyze different features of the image. Some deal with color, some with movement in specific directions, and some with lines and shapes. A "coordinating" region known as the primary visual cortex communicates signals to and from these areas. Recent brain scans using PET (Positron Emission Tomography), which registers where blood is flowing, support this description.

Researchers suggest that as the regions send and receive the nerve signals, so we see in our "mind's eye." Seeing and understanding may be one and the same process, and they seem to be an integral part of consciousness itself, at least for sighted people.

Light rays *coming into the eye are focused mainly by the dome-like cornea at the front of the eye, and partly by the lens a little way behind. Ciliary muscles around the lens make it fatter to focus on near objects and thinner for distant ones. In between the cornea and the lens is the pupil, the dark circle in the center of the colored iris. The pupil narrows in bright light to protect the retina and widens in dim light so that more light can enter.*

The retina contains the light-sensitive cells that send out nerve signals when light shines on them. The 120 million rod cells in each eye work well in bright or dim light, but see in monochrome (shades of gray).

The seven million cone cells concentrated around the back of the retina, where the main images fall, see details and colors, but function only in bright light. This is why evening scenes "gray out" – our cones cease working, and vision relies more on rods.

Insect eyes *are quite different from vertebrate eyes such as our own. Insects have compound eyes, made from a cluster of similar rod-shaped units known as ommatidia, each of which sees a fragment of the scene and gives the insect a mosaic view of the world. The more ommatidia there are in its eye, the better the insect sees.*

Worker ant eyes have just six ommatidia. The eyes of the female housefly shown here have 4,000. Dragonfly eyes have more than 30,000 to help them spot prey in mid-air. Few insects can see clearly anything much more than a yard away, but they are very good at spotting movement.

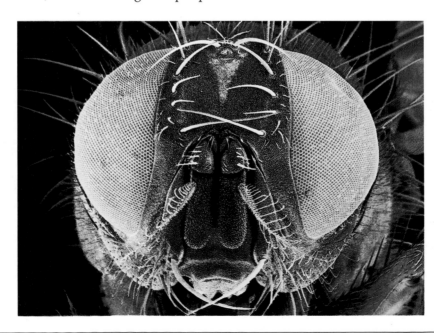

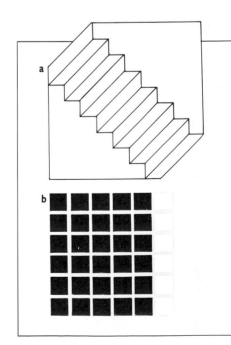

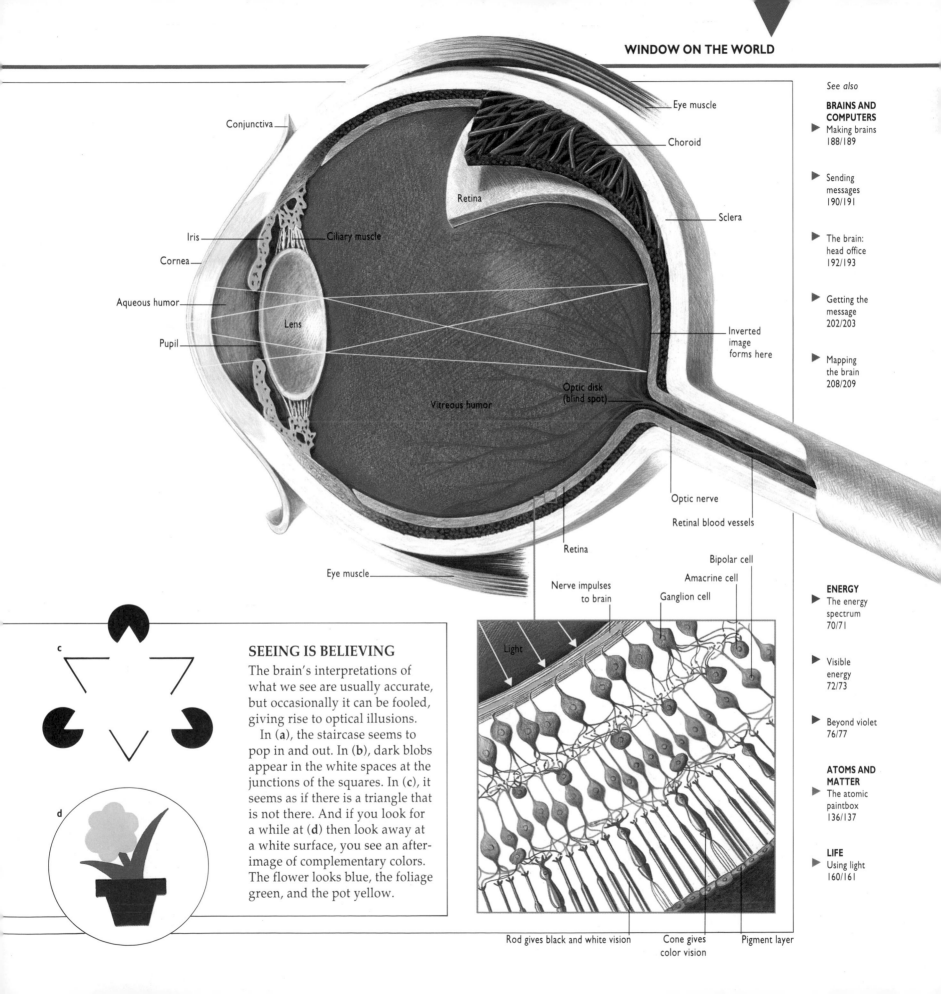

Conjunctiva

Eye muscle

Choroid

Retina

Iris — Ciliary muscle

Sclera

Cornea

Aqueous humor

Lens

Pupil

Inverted image forms here

Vitreous humor

Optic disk (blind spot)

Optic nerve

Retinal blood vessels

Retina

Eye muscle

Bipolar cell

Amacrine cell

Nerve impulses to brain

Ganglion cell

Light

Rod gives black and white vision

Cone gives color vision

Pigment layer

See also
BRAINS AND COMPUTERS
▶ Making brains 188/189

▶ Sending messages 190/191

▶ The brain: head office 192/193

▶ Getting the message 202/203

▶ Mapping the brain 208/209

ENERGY
▶ The energy spectrum 70/71

▶ Visible energy 72/73

▶ Beyond violet 76/77

ATOMS AND MATTER
▶ The atomic paintbox 136/137

LIFE
▶ Using light 160/161

SEEING IS BELIEVING

The brain's interpretations of what we see are usually accurate, but occasionally it can be fooled, giving rise to optical illusions.

In (**a**), the staircase seems to pop in and out. In (**b**), dark blobs appear in the white spaces at the junctions of the squares. In (**c**), it seems as if there is a triangle that is not there. And if you look for a while at (**d**) then look away at a white surface, you see an afterimage of complementary colors. The flower looks blue, the foliage green, and the pot yellow.

c

d

Sound sense

Although sounds are nothing more than subtle vibrations in the air, the ears detect them and the brain hears them as anything from traffic noise to music.

The ear flap on the outside of your head is just a small part of the ear. Inside is a tube that funnels sounds in and causes the eardrum to vibrate. The vibration of the eardrum, in turn, rattles three tiny bones called ossicles, sending pressure waves through the fluid of the inner ear.

The inner ear is a labyrinth of curling tubes and passages, but the key to hearing is the snail-shaped cochlea. This amazing structure is lined with a flexible skin called the basilar membrane which sits on rows of thousands of hair cells, each with 50–100 microscopic hairs. When the rattling of the ossicles sends pressure waves through the fluid filling the cochlea, the membrane waves about, too, playing over the hairs like hands plucking or pulling on harp strings.

The waving of the hairs triggers the hair cells into sending nerve impulses down the cochlear nerve to the brain. The cochlear nerve joins the auditory nerve which sends on the signals to the auditory (hearing) centers on the sides of the brain's cerebral cortex, where sounds are perceived and "heard" in our consciousness. New research has thrown up some surprising connections. It seems that in the split second before we become consciously aware of a sound's nature – whether from piano, guitar, human voice, road drill – the brain has already analyzed and identified its earliest vibrations. Also, what we think we hear, and especially how we perceive a sound's pitch, is closely related to the language and dialect we learn as a child. This has great implications for the understanding of music around the world.

The ear is sensitive to faint sounds because it amplifies the minutest vibration in the air. The outer ear amplifies slightly by funneling sounds toward the eardrum or tympanic membrane, tensioned by the tensor tympani muscle. But three linked bones – the hammer (malleus), anvil (incus), and stirrup (stapes), which together form the ossicles – are the ear's real amplification system.

The bones are connected and act as levers to increase the force of the vibration. The first bone in the chain, the hammer, moves a long way when rattled by the eardrum. The last, the stirrup, attached to the stapedius muscle, vibrates only a little way, but with more force. The

THE "SIXTH SENSE"

In addition to enabling us to hear, the inner ear also helps us to balance, with the aid of a labyrinth of passages called the vestibule – the three semicircular canals and the cavities of the utricle and saccule hidden just below the canals. The canals are aligned in three directions, so any movement of the head sets off currents in the fluid in at least one. The currents wave hairs rooted in sensory cells that fire nerve signals along the vestibular nerve to the brain. Slight differences in the signals sent from hair cells in the utricle and saccule in each ear help the brain sense position relative to gravity and the acceleration of head movements.

In addition, tiny spindle-shaped stretch-sensors (proprioceptors) embedded in muscles, tendons, and joints all over the body monitor the position of the head, neck, torso, and limbs. Information from the eyes reports the position and angle of the horizon.

With this information, the brain monitors body posture and balance. By way of the cerebellum, it moves muscles to help a gymnast balancing on a beam and to stop the rest of us from falling over.

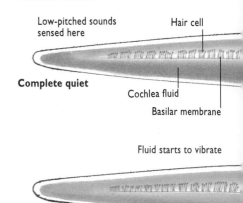

Uncurled cochlea

Low-pitched sounds sensed here

Hair cell

Complete quiet

Cochlea fluid

Basilar membrane

Fluid starts to vibrate

High-frequency sound

See also

BRAINS AND COMPUTERS
► Sending messages 190/191

► The brain: head office 192/193

► Getting the message 202/203

► The conscious mind 204/205

► Total recall 210/211

ENERGY
► What is pressure? 54/55

► Sound: vibrant energy 84/85

► Good vibrations 86/87

► Sound effects 88/89

LIFE
► Sound signals 162/163

sound is further amplified as the stapes knocks on the membrane covering the entrance to the inner ear called the oval window. The oval window is like the eardrum, but it is 30 times smaller, so the vibrations are compressed and intensified. More amplification depends on the ever-narrowing spiral of the cochlea.

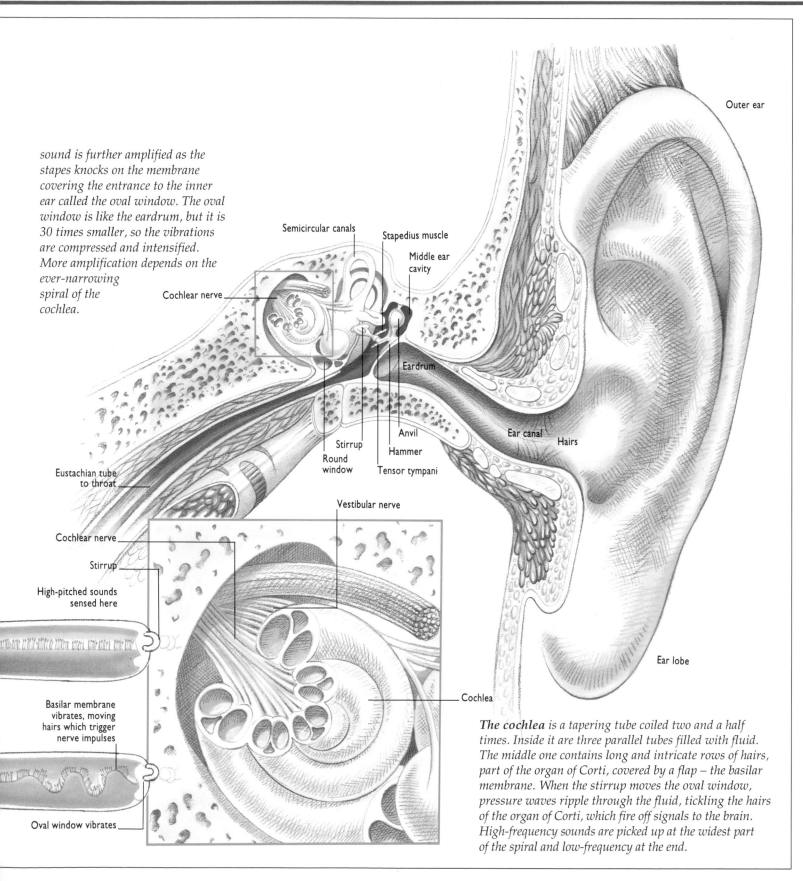

Semicircular canals

Stapedius muscle

Middle ear cavity

Cochlear nerve

Outer ear

Eardrum

Anvil

Stirrup

Hammer

Round window

Tensor tympani

Eustachian tube to throat

Vestibular nerve

Cochlear nerve

Stirrup

High-pitched sounds sensed here

Ear canal

Hairs

Basilar membrane vibrates, moving hairs which trigger nerve impulses

Oval window vibrates

Cochlea

Ear lobe

The cochlea is a tapering tube coiled two and a half times. Inside it are three parallel tubes filled with fluid. The middle one contains long and intricate rows of hairs, part of the organ of Corti, covered by a flap – the basilar membrane. When the stirrup moves the oval window, pressure waves ripple through the fluid, tickling the hairs of the organ of Corti, which fire off signals to the brain. High-frequency sounds are picked up at the widest part of the spiral and low-frequency at the end.

The chemical senses

Smell and taste respond to subtle chemical differences between substances.
In addition to helping us appreciate food, they can also affect our moods.

Have you noticed that when your nose is blocked by a cold, food has less odor – and less flavor, too? Smell and taste merge in conscious perception as they tell you about food and drink in your mouth. In fact, full appreciation of food requires these two senses, plus a third: the "temperature-texture" impression gained by the lips, mouth lining, and tongue. Smell and taste are known as chemosenses because their receptor cells respond to chemical substances. They may work by a "lock-and-key" effect, in which a certain shape of odor or flavor molecule fits into the same-shaped hole on the receptor cell, like a key in a lock firing off nerve signals.

Smell comes from a tiny patch of olfactory cells in the roof of the nose. There are probably between 6 and 30 different types of smell receptors here, with up to a million copies of each. Different chemical odors stimulate different combinations and numbers of receptors, and the olfactory centers in the brain probably perceive a smell from this pattern.

After reaching the brain's olfactory centers on the cortex, further signals pass out to other brain parts. These include the hippocampus and limbic system, which are basic and "primitive" parts that control moods and emotions, and also the thalamus. This is why the whiff of a long-forgotten place, person, or perfume can bring back deep-seated memories and feelings.

Taste comes from about 7,000 microscopic taste buds, most of them on the tongue. There are four basic taste qualities: sweet, salty, sour, and bitter. Sweet flavors are detected chiefly at the tongue's tip, salt along its front sides, sour along its rear sides, and bitter at the back. But the brain's taste centers may actually analyze the pattern of signals from all over the tongue.

As you breathe in, incoming air is diverted into your upper nasal cavity or olfactory cleft. As it eddies and swirls, airborne molecules settle on the olfactory epithelium and dissolve in mucus (right).

The olfactory epithelium is a thumbnail-sized patch containing several million slim sensory nerve cells. Each has a bunch of tiny hairs called cilia at its exposed tip and is surrounded by fatter supporting cells surfaced with microvilli (small fingerlike projections).

Different odors stimulate receptors on the cilia, generating nerve signals. These travel along the sensory cells' nerve fibers, which group into about 20 bundles. The bundles pass through holes in the skull bone that forms the nasal cavity roof to the olfactory bulb just above. The whole bulb responds to each smell, assesses the patterns of signals it receives from various patches of olfactory epithelium, and

THE SENSITIVE SHARK

Sharks track down their victims using a remarkable sense of smell. Odor particles enter through the nostrils and stimulate the olfactory sensors, linked directly to the olfactory lobes – almost the largest parts of the creature's entire brain! As the swimming shark's head moves, the concentration of odor varies. So this streamlined predator turns to the side on which the smell is strongest, until both sides are equal. In this way it can unerringly home in on its unfortunate victim from more than half a mile away.

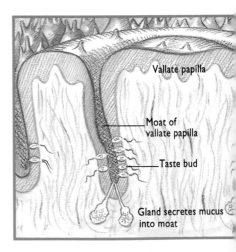

The "lumps" on the tongue are not the taste buds, but papillae. Vallate and fungiform papillae have taste buds in clefts. Vallate papillae at the rear of the tongue have most buds.

re-transmits signals to the brain along the olfactory nerve.

The sense of smell reacts more to change than steady stimuli. After a few seconds, smelling a smell, you have to inhale hard and concentrate to detect it.

Humans rely less on smell than other animals, but it still has survival value – detecting bad food, smoke or poisonous gases. It may also play a part in sexual attraction, through airborne hormones called pheromones which stimulate changes in behavior.

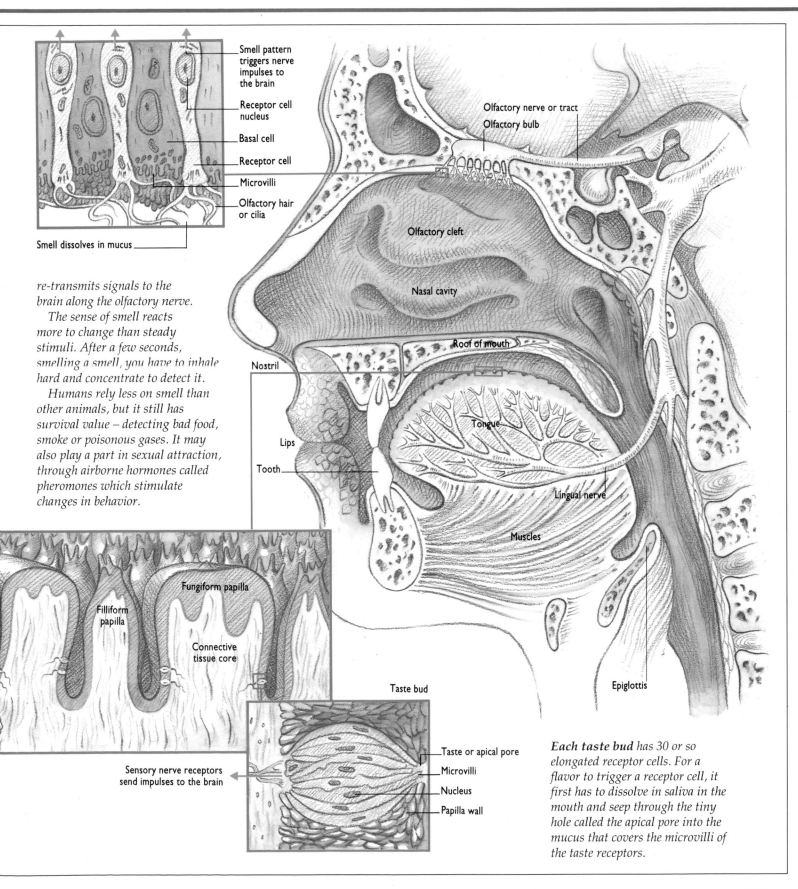

Smell pattern triggers nerve impulses to the brain

Receptor cell nucleus

Basal cell

Receptor cell

Microvilli

Olfactory hair or cilia

Smell dissolves in mucus

Olfactory nerve or tract

Olfactory bulb

Olfactory cleft

Nasal cavity

Roof of mouth

Nostril

Lips

Tooth

Tongue

Lingual nerve

Muscles

Epiglottis

Fungiform papilla

Filliform papilla

Connective tissue core

Sensory nerve receptors send impulses to the brain

Taste bud

Taste or apical pore

Microvilli

Nucleus

Papilla wall

Each taste bud has 30 or so elongated receptor cells. For a flavor to trigger a receptor cell, it first has to dissolve in saliva in the mouth and seep through the tiny hole called the apical pore into the mucus that covers the microvilli of the taste receptors.

Keeping in touch

We may not believe the evidence of our eyes or our ears, but when we touch something, then we know it is real. So how does the sense we can trust operate in the skin and the brain?

The human skin can detect and distinguish light pressure, heavy pressure, fast varying movements, and slow sustained ones, extra heat, loss of heat (cold), and pain ranging from discomfort to sheer agony – but the way in which it does so is not clear. The top layer of the skin is the epidermis, a layer of dead cells above living cells. Below this is the dermis, a thicker layer containing blood vessels, hair roots, sweat glands, and tiny, oddly shaped objects. These tiny objects are sensory receptors that send nerve signals to the brain when stimulated. But it is not certain how they work.

The skin seems to respond to four different kinds of sensation: heat and cold, a light touch, continuous pressure, and pain. However, it is hard to tell which receptors respond to each sensation. Some seem to respond chiefly to one kind of sensation, others to a variety. All four kinds of sensation are felt in places where the only receptors are little more than exposed nerve ends, while in other areas the receptors for each have their own shape. There are even differences in receptor arrangement between hairy and non-hairy skin.

It seems that the brain perceives touch according to the overall pattern of incoming nerve impulses. Nerves gather signals from millions of skin sensors over the body and convey them to the somatosensory cortex, the brain's touch center. Each patch of this cortex receives signals from one region of skin, such as the shoulder, arm, or finger. The size of each patch represents the sensitivity of its body part. So fingers have relatively large patches, the back has relatively small ones.

Experiments on monkeys have shown that the patches within the cortex can change in size with time. When monkeys are trained for tasks involving, say, sensitivity of the middle finger, the patch of cortex responsible for dealing with the middle finger's signals becomes enlarged. This is only one example of the brain's so-called plasticity. The brain is not a set of circuits etched on a board; it changes, modifies, and adapts throughout life.

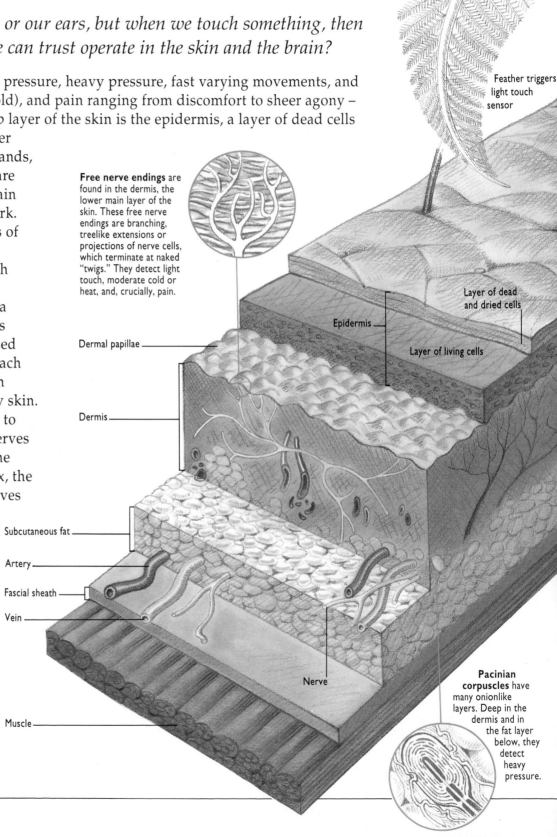

Feather triggers light touch sensor

Free nerve endings are found in the dermis, the lower main layer of the skin. These free nerve endings are branching, treelike extensions or projections of nerve cells, which terminate at naked "twigs." They detect light touch, moderate cold or heat, and, crucially, pain.

Layer of dead and dried cells

Epidermis

Layer of living cells

Dermal papillae

Dermis

Subcutaneous fat

Artery

Fascial sheath

Vein

Nerve

Muscle

Pacinian corpuscles have many onionlike layers. Deep in the dermis and in the fat layer below, they detect heavy pressure.

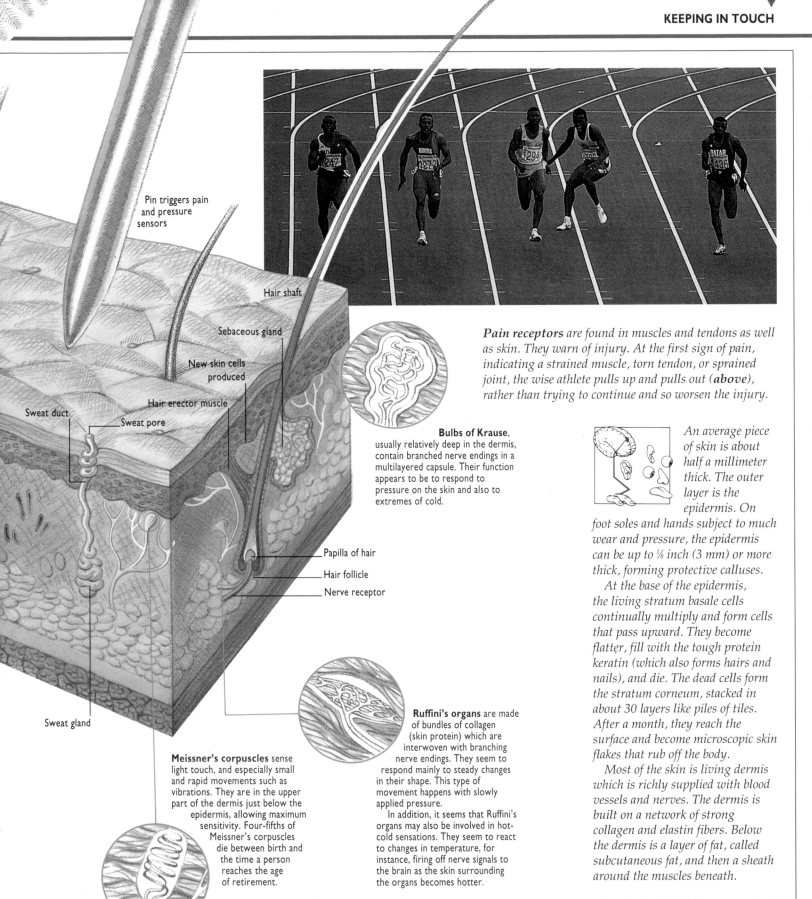

Pin triggers pain
and pressure
sensors

Hair shaft

Sebaceous gland

New skin cells
produced

Hair erector muscle

Sweat duct

Sweat pore

Papilla of hair

Hair follicle

Nerve receptor

Sweat gland

Pain receptors are found in muscles and tendons as well as skin. They warn of injury. At the first sign of pain, indicating a strained muscle, torn tendon, or sprained joint, the wise athlete pulls up and pulls out (**above**), rather than trying to continue and so worsen the injury.

Bulbs of Krause, usually relatively deep in the dermis, contain branched nerve endings in a multilayered capsule. Their function appears to be to respond to pressure on the skin and also to extremes of cold.

Meissner's corpuscles sense light touch, and especially small and rapid movements such as vibrations. They are in the upper part of the dermis just below the epidermis, allowing maximum sensitivity. Four-fifths of Meissner's corpuscles die between birth and the time a person reaches the age of retirement.

Ruffini's organs are made of bundles of collagen (skin protein) which are interwoven with branching nerve endings. They seem to respond mainly to steady changes in their shape. This type of movement happens with slowly applied pressure.

In addition, it seems that Ruffini's organs may be involved in hot-cold sensations. They seem to react to changes in temperature, for instance, firing off nerve signals to the brain as the skin surrounding the organs becomes hotter.

An average piece of skin is about half a millimeter thick. The outer layer is the epidermis. On foot soles and hands subject to much wear and pressure, the epidermis can be up to ⅛ inch (3 mm) or more thick, forming protective calluses.

At the base of the epidermis, the living stratum basale cells continually multiply and form cells that pass upward. They become flatter, fill with the tough protein keratin (which also forms hairs and nails), and die. The dead cells form the stratum corneum, stacked in about 30 layers like piles of tiles. After a month, they reach the surface and become microscopic skin flakes that rub off the body.

Most of the skin is living dermis which is richly supplied with blood vessels and nerves. The dermis is built on a network of strong collagen and elastin fibers. Below the dermis is a layer of fat, called subcutaneous fat, and then a sheath around the muscles beneath.

Getting the message

Throughout nature, organisms are constantly using their bodies and senses to broadcast and pick up messages vital to their existence.

Walk though a spring woodland and stop, look, and listen. There are birds chirping, butterflies dancing in pairs through a clearing, insects buzzing from flower to flower, and small creatures may be scurrying through the dead leaves underfoot.

The natural world is alive with animals sending and receiving messages. To do this, they may use light, sound and vibration, chemicals, touch or electricity. We see birds singing and displaying, dogs sniffing and licking. Even in the darkness of the deep sea, crabs and fish send and receive messages by waterborne chemicals and rippling currents.

What are they saying? Most animals communicate to satisfy three basic needs: for food, for living space or shelter, and to breed and raise young. Thus deep-sea fish may be luring curious creatures for prey, or warning competitors to retreat farther, or inviting mates to come nearer. Messages are relayed from sense organs to the creature's brain, giving information from which it can make its next actions: consume, attack, flee, ignore, mate, and so on.

In nature, certain basic messages are understood by all kinds of animals. For instance, a sound message, the hiss, is used as a defensive threat by creatures as diverse as beetles, toads, snakes, lizards, some birds, shrews, and cats. An often used visible message is the red-and-black or yellow-and-black warning coloration worn by certain caterpillars, wasps, bees, millipedes, starfish, salamanders, snakes, and birds. The colors usually mean: "Avoid me, I am poisonous/taste horrible."

Such "international" animal signals are called interspecific communication. They convey only simple messages. More subtle and complex are messages between members of the same species, called intraspecific communication. For instance, a foraging ant that finds a good food source lays an invisible "smell trail" back to the nest. The forager alerts its nest colleagues, who follow the trail to the food, which they then carry back to the nest.

The different sounds of chimps chattering among the trees reveal their moods and intentions, including contentment, dominance, hunger, anger, fear, terror, unease, aggression, and submission. The position of a wolf's ears, lips, teeth, and tail signal similar messages to other members of the pack. Vastly more complex and subtle is communication between humans.

Sight is immensely important to birds. They locate prey, predators, roosts, mates, nest sites, and nesting materials mainly by eye. One of the most dramatic visual signals in the whole animal kingdom is used by the peafowl. When courting, the peacock (**below**) erects and spreads his shimmering, multipatterned tail for the benefit of the peahen. She chooses the male with the most impressive overall display. His visual message says: "Choose me as a mate. I'm strong, healthy, and therefore fit to breed."

Grooming involves both sight and touch. There are at least two reasons why monkeys or apes in a troop groom each other. The first has to do with hygiene and health, the second with who bosses whom in the troop.

Dirty fur encourages pests and skin disease and lets in the weather. So the raking over and combing of fur and hair, especially in areas where individual animals find it hard to reach, and the removal of insect pests, is a very positive health measure.

But during grooming (**right**), the touching and stroking also help to establish and maintain social bonds and rank in the troop. Chief members are usually groomed first by subordinate members. Then the seniors may or may not groom the juniors, depending on the status of the juniors.

Some of the most exciting sound messages are sent by the great whales, and humpbacks (**below**) make a tremendous variety of clicks, screams, squeals, grunts, groans, and moans.

Researchers using underwater recorders and radio-tracking techniques are now beginning to understand how and why these amazing "songs" are produced. The reasons seem to be the age-old, basic ones. Male humpbacks appear to sing to warn off rival males, tell females they are available, and keep in touch in the vast oceans – the noises travel over 60 miles (100 km) because water transmits sound so well. Each whale has its own unique pattern of sounds, called a songprint, which it sings repeatedly to identify itself to far distant whales.

The conscious mind

The essence of mental awareness – your experience of the outer world and inner self – resides in your conscious mind.

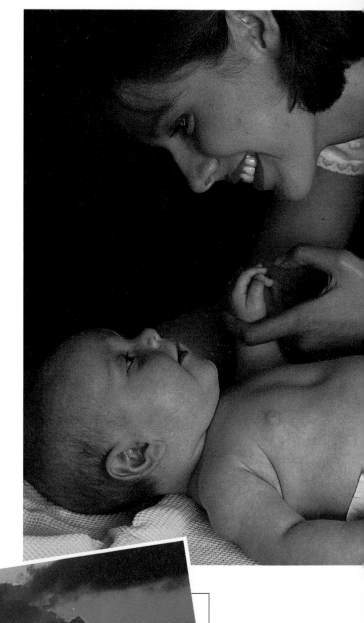

There are many answers to the question "Who are you?" You can answer simply by stating your name. Or you can think more deeply about what makes you an individual, including your personality, experiences, and memories. But whatever answer you give, the question is considered and dealt with in your conscious mind.

When you are conscious, you are aware of what is going on around you, thanks to all the billions of individual bits of neural information coming in from your sense organs. Vast amounts of filtering take place at a "lower" preconscious level, so that you attend only to the most significant signals. But being conscious can also mean being aware of your mind's internal thoughts and feelings, as well as your bodily movements and their results.

The human brain is so complex that scientists have yet to discover the specific sites and mechanisms of consciousness. But it seems that most of the brain is involved and that consciousness depends on nerve signals firing around vast and complex networks of nerve cells. Consciousness is also greatly affected by disturbances to the reticular formation, an area in the very core of the brain. Small changes caused by damage here can make a person either wide awake and aware, or drowsy and almost asleep. The reticular formation also has connections with the limbic system, which is involved in emotions and feelings.

Awareness may result, scientists argue, when nerve cells in the cortex, reticular formation, and other parts of the brain all fire nerve signals together, in a coordinated fashion. Another idea is that unconscious mental processes occur in the upper layers of the cortex, while conscious awareness involves nerve cells in the lower layers of the cortex.

IMAGINE THAT!

Look at a picture of an exotic island, and without much effort, you can picture yourself walking on the beach, lying in the sun, or splashing about in the warm water. To do this, your brain assembles sights from one place, smells from another, and so on, to build up the scene. To fill gaps, it makes lightning assumptions based on any similar, previous experiences. The result is imagination – not make-believe, but a mosaic of memories and informed guesses good enough to transport your consciousness to another time and place.

One of the key stages of a child's development is learning how to make representations of the world outside himself. This not only depends on the coordination of hands and eyes necessary to create a physical image on paper, but also on the child's awareness of himself as an individual.

The three-and-a-half-year-old child's image of his mother (**below**) shows that the child has learned to differentiate between body and head, and has an eye for detail: the mother is shown wearing glasses. The image is not an accurate representation, but there are still enough clues to enable the viewer's mind to identify it as a human figure.

*A **sense of self*** takes time to develop. In fact, young babies are unable to tell the difference between themselves and the rest of the world, including their mother (**above left**). The realization that there is a distinction between baby and the rest of existence develops between the ages of about six months and one year.

This nine-month-old boy (**above right**) is fascinated by his image in the mirror as he learns to identify himself as a separate individual, a learning process that appears to be almost unique to humans.

Most animals do not appear to have anything like our degree of self-awareness. A cat or dog that sees its reflection in a mirror may be temporarily affected and show fear, aggression, or curiosity. But on investigation, the animal in the mirror has none of the normal smells or sounds associated with its kind. The only animals thought to have even rudimentary self-awareness are our nearest relatives, the apes.

Sleep and dreams

No one can stay awake for more than a few days. But when you are asleep and your body relaxes, your brain switches on to dream.

Humans are daytime animals, with vision as the dominant sense. At night, there is little to do. Sleep probably evolved as a survival strategy for the dark hours. The body rests and repairs itself and stays still to save energy and avoid attracting attention.

But scientists find that the brain does not simply shut down during sleep. Using EEG (electroencephalograph) machines, they recorded the brain's electrical activity. They find startling changes through the night, especially during REM or rapid eye movement sleep, when the eyes flick about under closed lids. Four out of five people woken up during REM sleep say they are dreaming.

There are other changes during the night. The body moves its head, limbs, and position up to 30 times, to relieve the pressure on squashed nerves and blood vessels. Body temperature falls, as do heartbeat, breathing rates, blood pressure, urine production, and the levels of some hormones. The overall picture is of a relaxed body taken over by repair and maintenance processes.

Nearly all animals, from fish to gorillas, sleep. A predator such as the leopard can sleep deeply when it needs since it has little to fear. Dolphins have few predators, but must surface to breathe. So half their brain sleeps at a time, and they come up to breathe as needed. Prey animals such as antelopes are in constant danger so they snatch a few minutes' sleep here and there, and at a hint of trouble, they become alert.

Never seeming to sleep, the albatross (below) flies endlessly skimming over the oceans. Although it probably does not sleep in a sense that we would recognize, it is likely that it has moments when its brain is less active and it settles to repetitive body movements.

Dreams of flying (right) are quite common. Despite what some people believe, everybody dreams, but not everyone remembers their dreams when they wake up.

Some of the greatest thinkers throughout history have pondered on the meaning of remembered dreams. In The Interpretation of

Dreams *(1900), Sigmund Freud, the pioneer of psychoanalysis, proposed that dreams were the "royal road to the unconscious." He meant that dreams give insights to processes occurring and feelings being felt in parts of the mind that are not included in our conscious awareness.*

The body slips through several stages from wakefulness to deep sleep. It usually takes about 90 minutes to reach the first period of REM sleep, which lasts 10–15 minutes. Then the sleeper alternates between non-REM and REM over a 70–90 minute cycle. Most people have about four REM periods nightly. REMs lengthen and non-REMs shorten toward morning. We cannot usually recall dreams unless we wake during or just after REM sleep, when they occur. This is more likely to happen in the morning, hence vivid dreams occur if we doze off again just after waking.

Some researchers suggest that the day's experiences and memories are processed during sleep, especially REM sleep. In no particular order, the brain runs through the material, discards some, and files the rest as a strategy for survival. We inherit this essentially unconscious system from ancestral mammals, for whom basic survival was much more important. This is why dreams often do not impinge on, or make sense to, the conscious mind which humans have evolved.

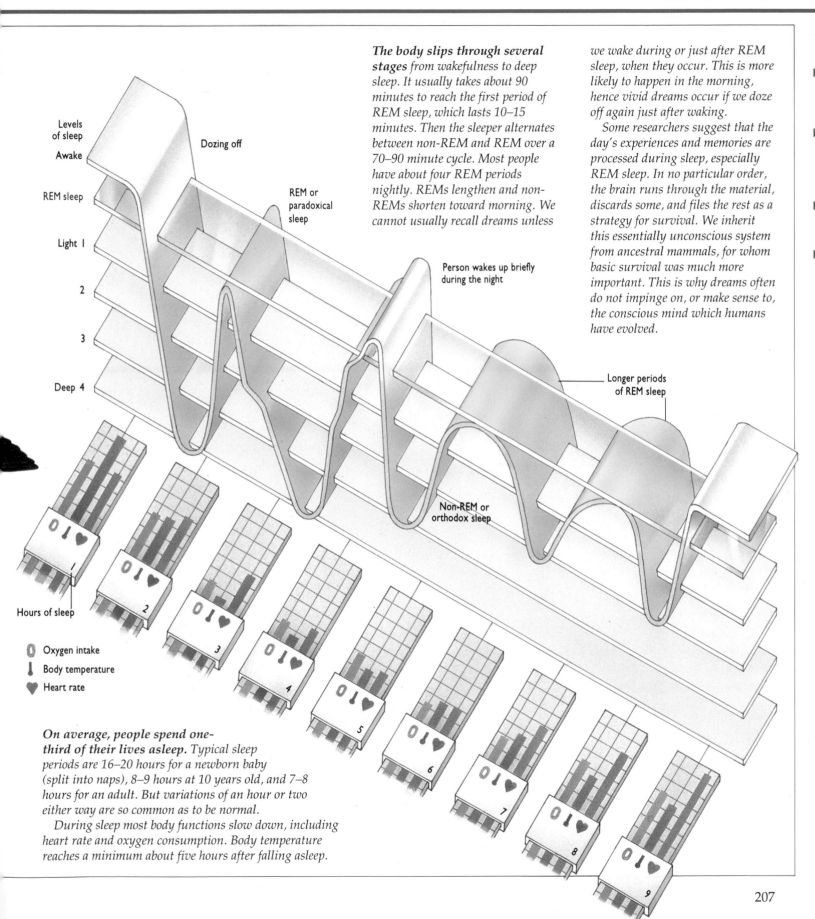

Levels of sleep

Awake

Dozing off

REM sleep

REM or paradoxical sleep

Light 1

Person wakes up briefly during the night

2

3

Deep 4

Longer periods of REM sleep

Non-REM or orthodox sleep

Hours of sleep

◐ Oxygen intake

♦ Body temperature

♥ Heart rate

On average, people spend one-third of their lives asleep. Typical sleep periods are 16–20 hours for a newborn baby (split into naps), 8–9 hours at 10 years old, and 7–8 hours for an adult. But variations of an hour or two either way are so common as to be normal.

During sleep most body functions slow down, including heart rate and oxygen consumption. Body temperature reaches a minimum about five hours after falling asleep.

Mapping the brain

Scientists are now using techniques that let them look inside the living brain to find out how it works.

Scientists know a great deal about the shape and structure of the brain, but are only now starting to understand more about which parts of the brain are responsible for such activities as thinking and learning. Traditional methods of finding out where the various brain functions are located involve studying people with brain damage or disease. But new scanning techniques are having a huge impact. PET scans (Positron Emission Tomography), for example, show blood moving in the brain, a good pointer to activity. Detailed images are made using MRI (Magnetic Resonance Imaging), which picks out types of brain tissues.

The newest of the scanners is SQUID, the Superconducting QUantum Interference Device, which records activity inside the brain. Whenever a neuron fires, it sends out a tiny electric current. This current creates its own magnetic field, which is picked up and mapped in minute detail by SQUID.

From traditional methods, it has been possible to find out something about which patches of the brain's gray matter have specialist roles. For instance, the left side of the brain largely controls actions on the right side of the body and vice versa. And in most people the left half of the cortex deals with language and logic, while the right handles non-verbal, visual,

*Hearing and seeing words are controlled by different parts of the brain. A PET scan reveals activity by recording the concentrations of blood flow in the brain. Inside a classroom (**above right**), spoken words trigger nerve impulses from the ears to the primary auditory center. Scientists had thought that written words were processed in the same way as sound. PET scans now reveal that written words are, in fact, picked up by the eyes and sent directly to the vision center, where they are processed.*

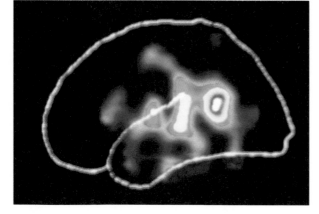

Hearing words

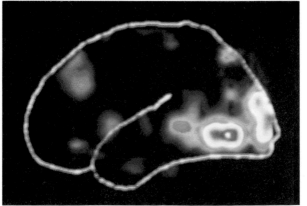

Seeing written words

and spatial skills. But the new scanners have more firmly identified many centers, including those dealing with the main senses of hearing, taste, and smell, as well as more "mental" functions such as speech recognition, written word recognition, word formation, and speech itself. For instance, as you read these words, a grape-sized lump of brain at the back of your head directly behind the eyes is signalling furiously. This is the primary visual cortex, or vision center. It assimilates and analyzes signals before passing them onto other cortex areas where concepts and ideas are understood.

Turning the page uses the motor cortex just under the area where the band of a set of headphones would sit; this area controls all movement. The motor cortex sends out signals along the motor nerves to the cerebellum below the cortex, and from there to the body's muscles, instructing them to move your body. Behind the motor cortex is the somatosensory cortex, sometimes called the touch or skin-sensation center. It receives nerve signals that bring information about touch, pressure, heat, and cold.

The cerebellum, constituting about one-ninth of the brain's total matter, is chiefly a relay and processing station for nerve signals sent from the cortex to the body's muscles and for signals coming in from the muscles and balance organs. When you learn a new skill, such as roller-skating, your early movements are jerky, clumsy, and uncoordinated. With practice, nerve links and signal pathways record the movements needed to instruct the correct muscles at the right times, making the movements smooth, precise, and "natural."

See also

BRAINS AND COMPUTERS
► Sending messages 190/191

► The brain: head office 192/193

► Window on the world 194/195

► Sound sense 196/197

► Keeping in touch 200/201

► Getting the message 202/203

ENERGY
► Sound: vibrant energy 84/85

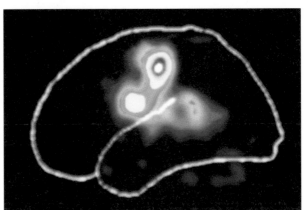

Thinking of words to speak

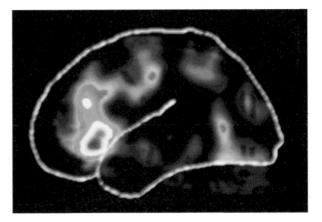

Speaking words

The processes of thinking about and then speaking a word make several areas of your brain start working. Thoughts are generated all over the brain, but the process of actually putting these thoughts into words is controlled by an area called the speech center.

When a word has been generated, messages are sent to the motor center, which coordinates the physical act of speaking. Signals are then sent out to the muscles of the voice box, neck, jaw, cheeks, tongue, and lips and the word is uttered.

Total recall

What's your name? Unless you are exceptionally forgetful, you know at once. But memory is not used only to recall simple facts.

When you see a picture, nerve signals from your eyes go to the visual centers in your brain, where you begin to analyze them. At once, memory comes into play. Quickly, you decipher these images and compare them with information stored in your brain. When a comparison fits, you visually recognize the object, and then a word-memory recalls the name of the object.

This happens in the working memory situated in the prefrontal lobes of the cerebral cortex, and it is vital for our ongoing view and awareness of what we sense, think, and do. Working memory is like the central processing unit of a computer, calling in data from the senses and longer-term memory stores, manipulating it, then "forgetting" almost at once as it moves on to the next set of tasks. The brain's shorter-term working memory can hold 10 or so items. Look up a phone number in the book, and you have to repeat it to yourself to "refresh" the memory as you make the call. If someone distracts you, the working memory is replaced with what they have said, and the phone number goes.

One vital memory structure is the hippocampus. It stores and sifts new information, for weeks or even months, then passes on significant parts of it for long-term storage. Longer-term memory, sited in parts of the cortex, holds facts, figures, past experiences, knowledge of language, mathematics, and so on.

Memory of an object, such as a pair of scissors, is probably not stored in one place. Information about scissors' shape, color, hardness, name, method of use, and function may each be stored in a different place. These pieces of data are brought together so that you can recall "scissors" in complete form when you hear the word, or see a pair or an image of them.

Many changes happen in the aging brain. These tend to affect shorter-term memory more than well-established longer-term memory. An older person may be unable to recall in precise detail what happened last week, yet can tell a story heard 50 years ago.

*Data in the longer-term memory is held in various parts of the cortex, possibly as sets of connected nerve cells. As new memories are laid down, connections and circuits develop and change as the brain's "wiring diagram" constantly adjusts and updates itself (**below**).*

In the brain's vast network of neurons, thin tendrils from the main "wire" or axon of each neuron send tiny electrical signals toward similar tendrils, the dendrites, of other neurons. In response to frequent nearby signals, from a repeated thought, perhaps, small bumps, called dendritic spines, develop on the dendrites.

Next, repeated electrical activity at the ends of the axons guides them to the dendritic spines, and a neural circuit associated with a "memory" becomes established. Over time, unused pathways may decay and wither, and the memory fades.

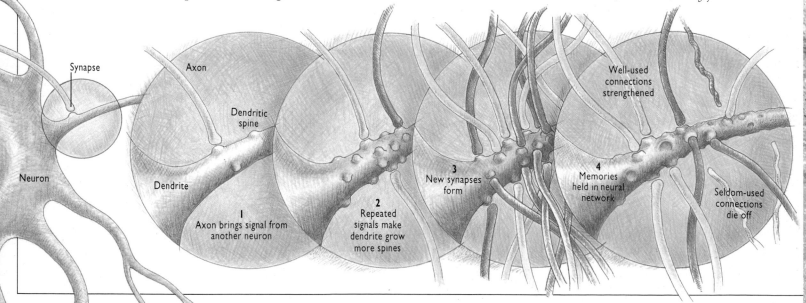

Synapse

Neuron

Axon

Dendritic spine

Dendrite

1
Axon brings signal from another neuron

2
Repeated signals make dendrite grow more spines

3
New synapses form

4
Memories held in neural network

Well-used connections strengthened

Seldom-used connections die off

MEMORY TEST

Look at the objects (**above**) for 20 seconds, close the book, and see how many you can recall 20 seconds later. If it is not many, there are techniques that help.

Visualize the whole picture, with all objects in place. Try to retain the overall positions as a photo "snapshot"; or name each object and arrange the initial letters into a word that you remember easily; or make up a story about the objects so that when you recall the first one or two, the rest follow.

The higher functions

Only humans, it seems, can learn lessons, make decisions, solve hard problems, predict outcomes, level judgments, create ideas, and communicate using language.

The brain's "higher functions" are vastly more sophisticated and complex than the low-level functions such as reflexes and simple stimulus-response reactions. For example, a bird spots food in a garden bird-feeder. It pecks at random, sometimes for hours, until by trial and error it hits the right spot and gets the food. A human would look at the feeder, understand how it works, and devise an action to trigger the feeder mechanism before taking any physical action.

The higher functions seem to be located in the cortex, the outer gray-matter coat of the cerebral hemispheres. Specific cortex areas deal with data from sense organs or with signals going out to muscles. But higher functions seem more widespread and diffuse. They appear to be interlinked activities taking place in many parts of the brain. Despite their awesome complexity and sophistication, higher functions still seem to rely on enormous networks of nerve cells making connections and flashing signals between each other.

Experiments show that the brain processes information in hierarchies. Like the many workers on the factory floor, and the few bosses in the office, some nerve-cell networks deal with small details while others handle more general, higher-level information. For instance, when you recognize the face of an old friend, your eyes pass nerve signals through visual pathways in the base of your brain. The details of the face, such as its colors, lines, shapes, and movements, are processed separately, in different parts of the visual system near the back of the brain. The results are fed to a lower part of the temporal lobe on the side of the brain. Here, the information comes together and activates relatively few "face cells," whose activity represents your

Reading shows the brain's power. *Written or printed marks on a piece of paper are effortlessly translated into a series of ideas and images. One person's thoughts, recorded days, months, years, or even centuries ago, can be passed on and the ideas used to educate, advise or entertain.*

Reading demonstrates our power *to think in the abstract. This means thinking about objects and ideas that are not being directly experienced.*
Learning to read is one of the most important steps in a person's life – it opens the door to the wealth of written material. Learning a foreign language and reading in it shows the mind's power even more clearly.

A remarkable feature of human thought is the ability to project situations and events into the future and to plan ahead in the mind, entirely in abstract terms. This is true for all manner of problems, from what to buy on a shopping trip to how to construct a skyscraper (**left**). After the initial thoughts, the next stage is to visualize and record the plan. This is done by writing down a shopping list or drawing blueprints.

The fact that humans can cooperate on complicated and long-drawn-out tasks is one of the reasons why we have come to dominate the animal kingdom and the physical environment of the planet. Our higher functions have proved to be an evolutionary winner.

Humans are highly visual animals. In fact, we often think in pictures, and we readily learn to understand abstracted, simplified images such as symbols on a map (**above**). Although the symbol for a church hardly looks like a real church, we can transfer the meaning from a shape on the map to a building in the landscape. But where we really score is in our ability to use all of our functions, virtually at once. We do not have to switch consciously from one to another; they are all available all the time.

recognition. These fire off new sets of nerve-cell pathways involving the language and speech systems, so you say "Hi!"

No other creatures have our abilities with language. Words, the basic units, are assembled into phrases and sentences when you speak. Meanings of words exist as concepts of "ideas" in nerve-cell networks across large areas of both the left and right cerebral hemispheres. Actual words, along with information on grammar and sentence formation, are in another system of nerve networks, mainly in the left hemisphere. A third system, also on the left, acts as an intermediate, translating the chosen concept into the relevant words.

Thinking in numbers

Numbers play a crucial role in modern life. They are used in everything from ordering food to understanding the universe.

Numbers are an answer to the question, "how many?" In early human societies they were used only in relation to the real objects of everyday existence – how many cows, how many houses, and so on – and there was no need for more than a few numbers. Any large number could be summed up simply as "many."

Numbers were spoken and indicated with fingers, stones, and even piles of shells for thousands of years before they were written down. People learned to make permanent records of numbers in countless ways, from the knots on cords called quipus used by the Incas to the notches on tally sticks used by British tax collectors right up until 1828.

Written numbers or numerals were probably invented about 5,000 years ago when people first began to live in towns. The oldest known numeral systems are those of the ancient Egyptians and the Sumerians.

In the west, written numerals were only used for drawing up a calendar or codifying the power of a ruler – or for the musings of philosophers and thinkers. Indeed, many Greeks regarded numbers as the sacred – and secret –

The Chinese abacus was the earliest mechanical device to aid reckoning and could be used to tally numbers beyond a billion. The values of beads increase by ten times on each successive wire, and beads to the right of the bar are worth five times beads to the left of the bar. Numbers are shown by sliding beads toward the bar.

NUMBERS GAME

Over the course of history many different civilizations developed and used their own number systems. Each of the examples of different systems in action (**right**) demonstrates how one system depicts and works out to be the number 11,416.

With its unique symbols and simple rules of use, the Arabic, or decimal, system, has come to dominate. With it numbers can be multiplied and divided simply – something almost impossible when using, say, the Sumerian or the Roman systems.

11,416

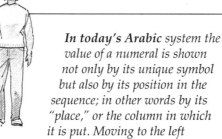

In today's Arabic system the value of a numeral is shown not only by its unique symbol but also by its position in the sequence; in other words by its "place," or the column in which it is put. Moving to the left multiplies the value of the symbol by 10 for every move. Thus the 1 in 16 equals ten, the 4 in 416 equals four hundred, the first 1 in 1,416 equals one thousand and so on. Adding all the values gives 11,416.

| 1 | 2 | 3 | 4 | 5 | 6 | 7 | 8 | 9 | 10 |

10,000	1,000	100	10	1
1	**1**	**4**	**1**	**6**
10,000 +	1,000 +	400 +	10 +	6

preserve of priests. Only gradually did numerals find their way into the everyday world, and some people still find them hard to deal with.

A great variety of ways of writing down numbers has been devised since the earliest systems, but the one used almost universally today is the Arabic system. In it, numbers are arranged essentially in groups of 1 and 10, which is why it is called the decimal system, from the Latin *decima* meaning 10.

The Arabic system is based on a system developed from the Hindu one by Arabic scholars such as al-Khwarizmi about 1,300 years ago. Arabic numerals are now so familiar that they seem natural to us. Yet it was not until the 13th century that this system of numerals spread into Europe – and it took over three centuries for it to displace the system of Roman numerals then in use.

Nowadays, we often think of numbers in abstract terms, unconnected with real objects, and our number sequence runs from zero right to infinity. We have words and ways of noting down numbers that are much too large to count physically, such as a million, a billion, and even a "googol," which is 1 followed by 100 zeros. We are even familiar with the idea of negative numbers – that is, numbers that have a value less than zero.

HIGHER POWERS

Very large numbers and fractions can be written compactly in multiples of ten. Thus 100 is 10^2 since it is 10 x 10, or ten to the power two; 1,000 is 10^3 or ten to the power three; a million is 10^6, a billion is 10^9. Using this method a number such as 3,600,000 can be written as 3.6×10^6 ($3.6 \times 1,000,000$).

Numbers less than 1 can also be written in this compact way. For example, 10^{-3} represents 1 divided by 10^3, or one thousandth (0.001), so 3.6×10^{-6} represents 0.0000036.

$$10^2 = 10 \times 10 = 100$$

$$10^3 = 10 \times 10 \times 10 = 1,000$$

$$10^{-3} = \frac{1}{10 \times 10 \times 10} = 0.001$$

$$3.6 \times 10^6 = 3,600,000$$

$$3.6 \times 10^{-6} = 0.0000036$$

See also

BRAINS AND COMPUTERS
▶ The higher functions 212/213

▶ Calculating machines 216/217

▶ Codes for computing 218/219

SPACE
▶ Creation of the cosmos 42/43

ENERGY
▶ The energy spectrum 70/71

ATOMS AND MATTER
▶ Exotic particles 144/145

The Sumerians used two symbols, representing 1 and 10. One 10 symbol plus six 1 symbols means 16. Placing a symbol farther to the left multiplies its value by 60. So the 10 symbol to the left of the 16 means 600 – that is, 60 x 10. The three 1 symbols to the left of the 600 are 3 x 60 x 60 which equals 10,800.

x 60 x 60	x 60	1
⋁⋁⋁	◀	◀⋁⋁⋁
10,800 +	600 +	16

Roman numerals – groups of symbols: I for 1, V for 5, X for 10, and so on – were used to keep records rather than make calculations, which were performed on a grooved board.

I	II	III	IV	V	VI	VII	VIII	IX
1	2	3	4	5	6	7	8	9

X	L	C	D	(I)	((I))
10	50	100	500	1,000	10,000

10,000	1,000	100	10	5	1
((I))	(I)	CCCC	X	V	I

| 10,000 | +1,000 | + 400 | + 10 | +5 | + 1 |

The Maya of Central America used a place-value system based on 20. Each numeral was made up of dots, for 1, and lines, for 5, written in columns. Each line in the column was further multiplied by 20.

● 1   ——— 5

x 8,000	●	8,000
		+
x 400	●●●	3,200
		+
x 20	═══	200
		+
1	● / ≡	16

Ancient China had many systems. In this one, sticks were arranged in squares on a counting board. Numerals increased in value tenfold if set one square farther to the left.

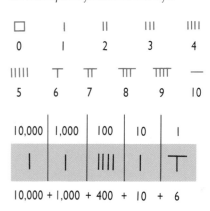

□	I	II	III	IIII
0	1	2	3	4

IIIII	丁	丅丅	丅丅丅	丅丅丅丅	—
5	6	7	8	9	10

10,000	1,000	100	10	1
I	I	IIII	I	丅

| 10,000 + 1,000 + 400 + 10 + 6 |

Calculating machines

So important is the need to keep count and calculate at speed that arithmetic was mechanized as soon as people had worked out how it could be done.

Until the 17th century, the only things that people had to help them calculate were the abacus and simple tables of mathematical quantities, such as fractions. Tables spared mathematicians the job of repeating laborious calculations, and the abacus – an extremely efficient and speedy device – was found wherever there was trade. Where the abacus has been traditionally used, it has only slowly been displaced by the pocket calculator in the 20th century.

Back in 1617, however, Scottish mathematician John Napier invented a calculation aid called "Napier's bones." These were a set of moveable multiplication tables engraved on ivory rods. But Napier had already made a far more important

innovation in 1614 when, after years of work, he published the first table of logarithms. These reduced multiplication and division to simple addition and subtraction. William Oughtred, an English clergyman and mathematician, devised a mechanical aid to calculation based on Napier's logarithms. It consisted of scales of numbers marked on twin rulers. Each number was marked at a position corresponding to its logarithm. Multiplying and dividing numbers became a matter of sliding one scale against the other. This developed into the slide rule, used by scientists and engineers up until the 1970s.

A clever mechanical calculator of the 19th century was the Difference Engine, invented by Englishman Charles Babbage.

The first cogged-wheel calculator (right) was built in 1642 by the French scientist, philosopher, and mystic, Blaise Pascal, to aid the work of his father, a tax collector. It could only add and subtract.

Called the Pascaline, it used cogged wheels which the operator turned with a stylus to add numbers. Numbers were displayed on cylinders seen through holes above the wheels. When a cylinder turned past the 9 position, its neighbor on the left turned by one division.

The first was finished by Babbage in 1822, and several were used for astronomical and life-insurance calculations.

The most advanced mechanical computer devised was Babbage's Analytical Engine. It embodied the essential features of modern computers. Numbers were to be input on punched cards, kept in the machine's "store" or memory, processed in an arithmetical section called the "mill," and printed out. Sadly, the machine was never completed.

The first machine to add and print results was made by American William S. Burroughs in 1890. George R. Stibitz made the first electromechanical calculator in 1940. But mechanical and electromechanical calculating machines were soon to be outmoded by the electronic calculator.

See also

BRAINS AND COMPUTERS
▶ Thinking in numbers
214/215

▶ Codes for computing
218/219

▶ The silicon revolution
220/221

▶ The number crunchers
222/223

▶ Modeling reality
226/227

ANALOG VERSUS DIGITAL

Calculating devices such as the abacus and the pocket calculator are digital – they represent numbers by means of a series of numerals. But slide rules and traditional clock faces are analog devices, representing numbers on continuous scales.

The term "analog" is used because such devices work using processes that are a "picture," or analog, of what the calculations are being made about. Thus the slide rule adds and subtracts lengths, which is an analog of adding and subtracting logarithms. The movement of a clock's hour hand is an analog of the Sun's revolution around the sky.

The most advanced analog mechanical computer ever built was Vannevar Bush's Differential Analyzer, in which the rotation of wheels represented mathematical quantities. It was built at the Massachusetts Institute of Technology in 1931.

The vast majority of today's computers are digital. But analog electronic computers are used for certain jobs where near-instant response is vital, such as in flight simulators (**below**) used to train pilots to fly today's complex aircraft.

In flight simulators, direction and speed of flight, the forces acting on the aircraft, and engine response are all "modeled" by voltages and currents in electronic circuits. These circuits *then control the readings of the instruments on the flight deck, the computer-generated scene through the windows, and the response of the throttle and pedals used by the aircrew.*

Codes for computing

Using simple codes and logic systems, networks of switches are turned into the remarkable electronic "brain" of the computer.

Computers have become so sophisticated, fast and powerful that it is sometimes easy to forget that they are in essence quite simple machines based on simple principles. Nearly all a computer's operations are based on electronic switches that can only be on or off. A computer's ability to perform complex operations depends on the way these switches are linked together and used to open and shut different circuits.

The secret lies in a system of numbers called the binary system. In this system, numbers go up in multiples of two, not ten as in the decimal system we use every day. So binary numbers can be written using just two symbols instead of ten. These two symbols are usually 1 and 0, but could just as easily be yes and no, black and white, on and off, dot and dash – that is, any pair of alternatives. In a computer, a single digit in the binary system can be an electronic switch which may be either on or off. So numbers are built up from combinations of binary digits, or bits, for short.

Bits are the basis of all computers. Not only numbers, but all kinds of data, including letters, pictures, and even sounds, are built up from combinations of bits, each representing a different letter, picture instruction, and so on, in coded form. Under the widely used ASCII code, for instance, each letter of the alphabet is represented by seven bits. The code for the letter A, for instance, is 1100001; the letter B is 1100010; C is 1100011; and so on.

Not only does a computer represent data, but it also processes it in two ways. Simple calculations rely on binary arithmetic as shown below; complex tasks

Binary numbers are written as multiples of two, not ten as in decimals. Decimal numbers are written in reverse order in units, tens, hundreds, and so on. Binary numbers are written in reverse order in units, twos, fours, eights, and so on. Decimal 5, for instance, is binary 0101: 0×8, 1×4, 0×2, 1×1 ($0 + 4 + 0 + 1 = 5$).

	8	4	2	1
0				
1				●
2			●	
3			●	●
4		●		
5		●		●
6		●	●	
7		●	●	●
8	●			
9	●			●
10	●		●	

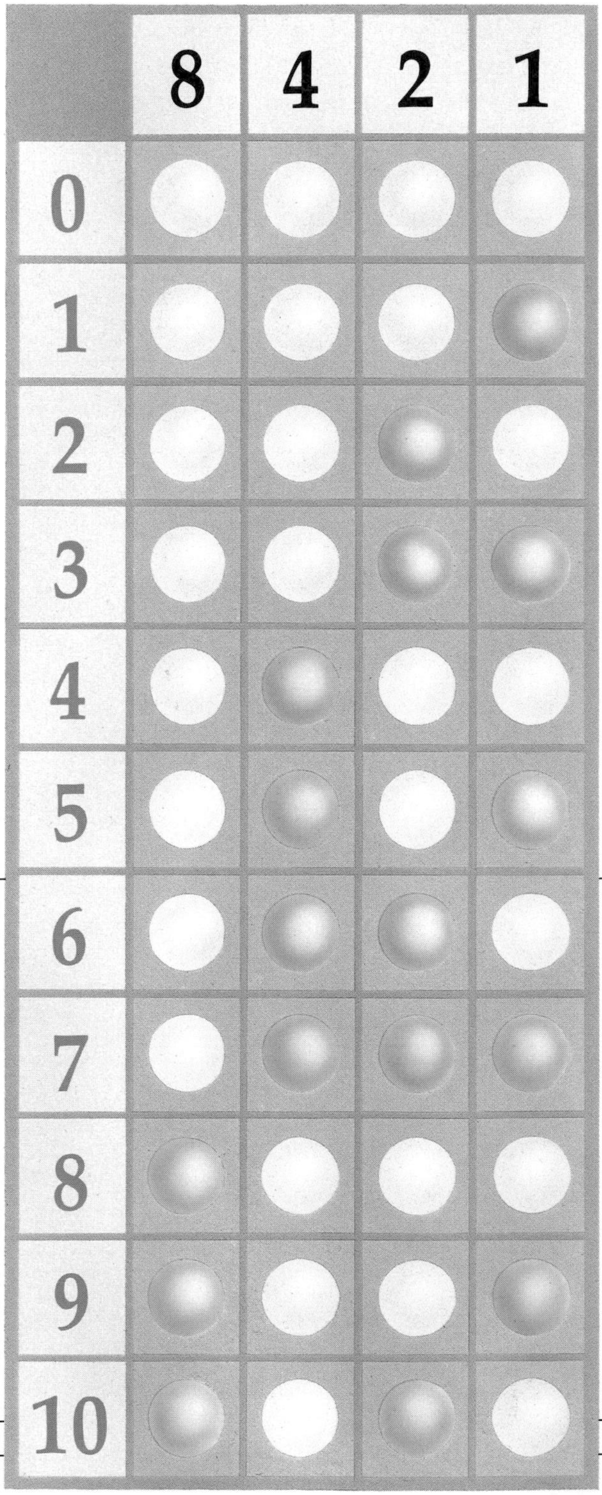

BASIC BINARY

Computers reduce all mathematical operations to the simplest possible form – counting. They add, subtract, multiply, and divide numbers in much the same way as a child might use counters – only they do everything in tiny fractions of a second.

Every binary number is added or subtracted just as decimals are, with digits carried over to the next place when one place is full. In other words, whenever there are two 1s in the same place, a 1 is carried over to the next place. Thus 0 + 0 equals 0, 0 + 1 equals 01 (decimal 1), but 1 + 1 equals 10 (decimal 2). Similarly 10011 (decimal 19) added to 01100 (decimal 12) equals 11111 (decimal 31).

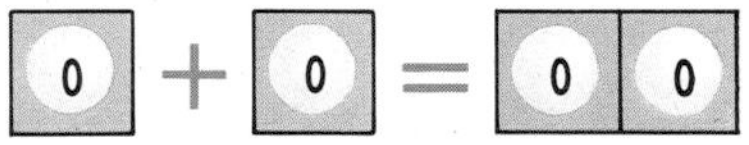

| 0 | + | 0 | = | 0 | 0 |

0 plus 0 equals 0 carry 0

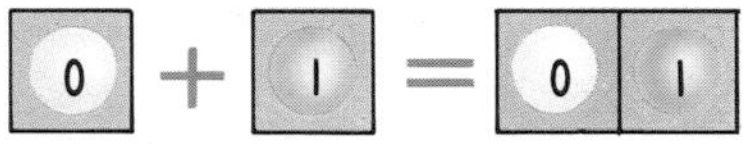

| 0 | + | 1 | = | 0 | 1 |

0 plus 1 equals 1 carry 0

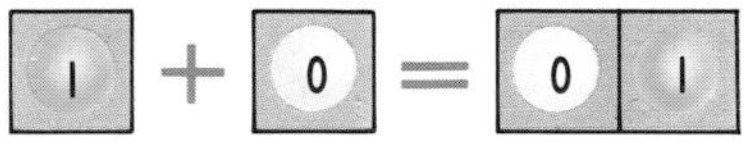

| 1 | + | 0 | = | 0 | 1 |

1 plus 0 equals 1 carry 0

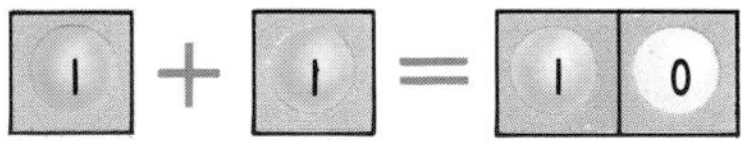

| 1 | + | 1 | = | 1 | 0 |

1 plus 1 equals 0 carry 1

What's the point? The flow of data in a computer is directed by opening and closing switches, rather as trains are directed one way or the other by the switching of points. Just as a train's destination depends on which points are open, so the outcome of a computation depends simply on which circuits are switched on.

depend on "logic" systems first suggested in the mid-1800s by English mathematician George Boole (1815–64). This system of logic was a vital element in the leap from simple calculating machines to processing computers achieved in the 1940s.

The switches and circuits in a computer can be thought of as gates that open or shut in response to various bits of data according to simple logic, which is why they are called logic gates. The four most common types are AND, NOT, OR, and E(Exclusive)OR gates. An AND gate opens and current flows if one switch in the circuit AND all the others are on. A NOT gate opens if one switch, but NOT the others, is on. An OR gate opens if one switch OR any switch is on. An EOR gate opens if only one switch is on. By connecting logic gates together so that the output of one controls the input of another, computer circuitry can be made to perform complex data processing tasks.

Adding two numbers in binary is like decimal addition, but carrying when there are two ones or more in the column instead of a ten or more. With 7 (0111) plus 3 (0011), if you add the first column together, the two 1s gives a 0, carry 1. In the next column there are now three 1s so the answer is 1 carry 1. In the third column there are now two 1s so a 0 is written down and a 1 carried to give the result 1010, or decimal 10.

Multiplication is reduced to a simple addition. In binary multiplication of 7 by 3, the end result is 10101, which is decimal 21.

$7 + 3 = 10$

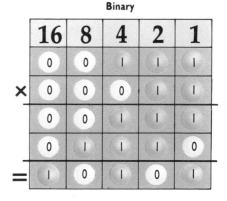

$$7 \times 3 = 21$$

The silicon revolution

Thanks to the tiny silicon chip, huge computing power and sophisticated control systems can be packed into tiny computers and computer-controlled devices.

In the early computers of the 1940s, the electronic switches that handled and processed all the data were thousands of glass valves, which looked a bit like electric light bulbs. But even low-powered valve computers were gigantic, the valves generated much heat and they and other high-voltage components frequently burned out. Only when valves were superseded by tiny transistors in the 1950s could the computer revolution really begin.

Transistors depend on materials called semiconductors, such as germanium and silicon, which conduct electricity only partially. When electrodes are implanted in solid lumps of these materials, they can be made either to transmit a current or to block it, and so act as a switch.

Single transistors are tiny enough – they are rarely much bigger than a pea. But modern computers and electronic devices exploit the power of the silicon chip, or microchip, in which the connections for numerous transistors are etched onto a single chip or wafer of silicon just a few millimeters across. In the simplest microchips, like those in personal stereos, there may be no more than a dozen transistors, but the remarkable microprocessors inside computers are single chips containing a million or more transistors, all linked in a single integrated circuit.

The appearance of microprocessors in 1971 started a revolution in computing because complex circuits could be built into a tiny space and reproduced very quickly and cheaply. With the microprocessor came the first micro computers, computers small enough to fit on a desk or even in a pocket. Nowadays, the existence of microprocessors means that highly sophisticated computer control circuits are found not only in computers, but also in countless ordinary electronic devices including pocket calculators, cameras, watches, stereos, and TV sets.

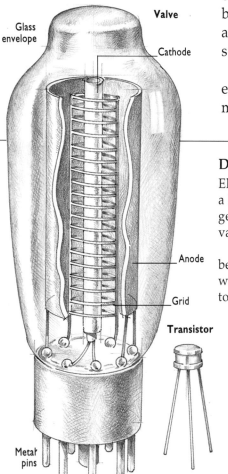

Glass envelope

Valve

Cathode

Anode

Grid

Transistor

Metal pins

DOWN THE GENERATIONS

Electronic computers have come a long way since the first generation, based on triode valves, and shrunk dramatically.

The triode valve worked because current flows only one way – from the cathode filament to the surrounding metal anode. The metal grid in between is a third electrode – which is why it is called a tri-ode – and this controls the current.

Valve computers filled entire rooms and fans were fitted to carry away the heat. Even so, valves had only short lives, needing frequent replacement.

The invention of the triode transistor in 1948 ushered in the smaller, more robust, second generation of computers. The integration of several transistors into a single microchip gave birth to the third generation. The fourth generation came with very large-scale integration (VLSI) in the 1970s, when hundreds of thousands of components were squeezed into a tiny microchip.

The latest, fifth generation, links all these processors in parallel for phenomenal computing speeds.

Inside a computer the various microchips for memory and processing are laid out on the printed circuit board or PCB and connected by a network of "tracks" or wires printed in copper. Each chip is a tiny but fantastically complex integrated circuit or IC etched on a silicon chip and consisting of millions of individual transistors, which are essentially electronic switches capable of being either "on" or "off." The chip is encased in plastic or ceramic and linked via gold wires to metal pins connected into the circuit board.

The chip's architecture is rather like the circuit board in miniature, and each has its own central processing unit, memory areas, input/output controller, and so on.

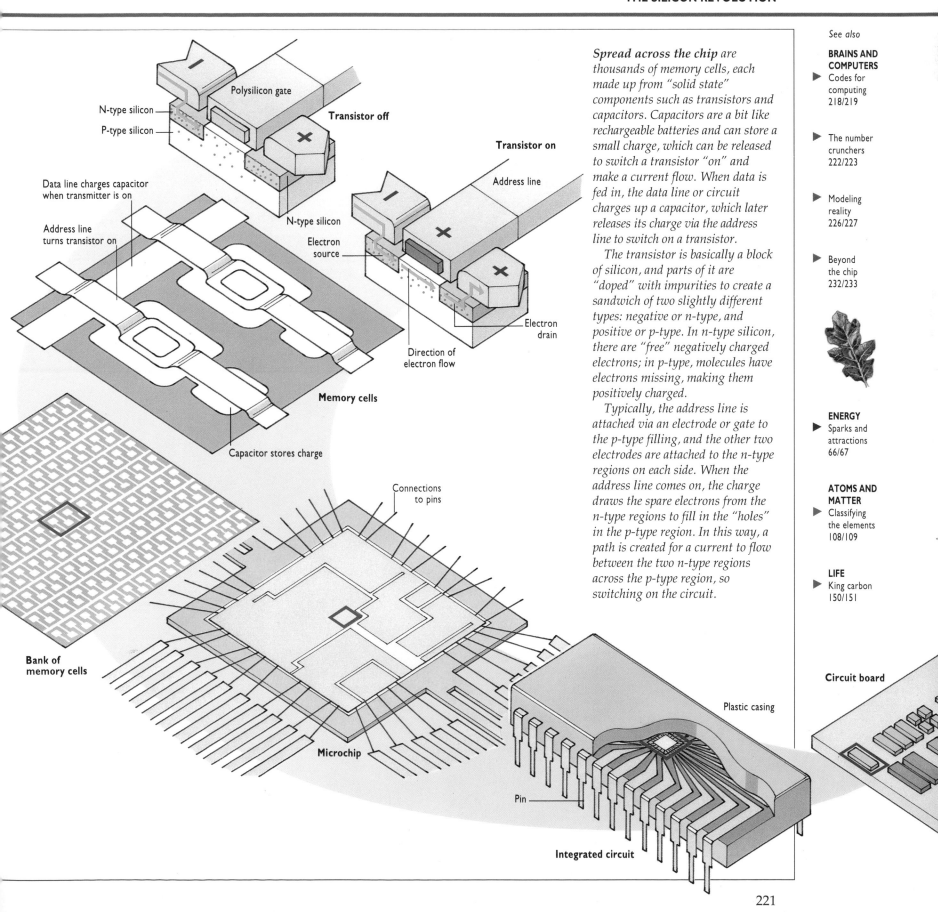

N-type silicon

P-type silicon

Polysilicon gate

Transistor off

Transistor on

Address line

N-type silicon

Electron source

Data line charges capacitor when transmitter is on

Address line turns transistor on

Electron drain

Direction of electron flow

Memory cells

Capacitor stores charge

Connections to pins

Bank of memory cells

Plastic casing

Microchip

Pin

Integrated circuit

Circuit board

Spread across the chip are thousands of memory cells, each made up from "solid state" components such as transistors and capacitors. Capacitors are a bit like rechargeable batteries and can store a small charge, which can be released to switch a transistor "on" and make a current flow. When data is fed in, the data line or circuit charges up a capacitor, which later releases its charge via the address line to switch on a transistor.

The transistor is basically a block of silicon, and parts of it are "doped" with impurities to create a sandwich of two slightly different types: negative or n-type, and positive or p-type. In n-type silicon, there are "free" negatively charged electrons; in p-type, molecules have electrons missing, making them positively charged.

Typically, the address line is attached via an electrode or gate to the p-type filling, and the other two electrodes are attached to the n-type regions on each side. When the address line comes on, the charge draws the spare electrons from the n-type regions to fill in the "holes" in the p-type region. In this way, a path is created for a current to flow between the two n-type regions across the p-type region, so switching on the circuit.

See also

BRAINS AND COMPUTERS
▶ Codes for computing 218/219

▶ The number crunchers 222/223

▶ Modeling reality 226/227

▶ Beyond the chip 232/233

ENERGY
▶ Sparks and attractions 66/67

ATOMS AND MATTER
▶ Classifying the elements 108/109

LIFE
▶ King carbon 150/151

221

The number crunchers

As computers have become more and more complex and sophisticated, so they have become increasingly easy to use.

Programmers and computer designers have put such effort into simplifying the interface between user and computer that the operator is barely aware of the whirlwind of electronic activity inside the computer as every command is executed.

Whenever you press a key on the keyboard, a program continuously scanning the keyboard to detect key contact instantly generates the right sequence of bits, the binary code of zeros and ones with which the computer works. The sequence or "word" is instantly dispatched to the computer's central processor unit which stores it in one of a series of registers where it can then be manipulated electronically. It is then given an "address" to make sure it goes to the right cell of the computer's memory.

At the same time, other programs stored in microchips within the computer, or on the hard disk or floppy disks, perform countless tasks, including managing the display screen or printer and controlling the storage and retrieval of information on the disks.

Inside a computer *are a series of microchip circuits. One is the central processing unit (CPU), which is at the heart of the computing activity. It contains the arithmetic and logic unit where calculations and logical operations are carried out; the control unit which decodes, controls, and executes programs; and the immediate-access memory which stores all the data and programs the computer is currently working with. All three work in units called registers, which are special memory locations designed for instant access.*

There are two basic kinds of memory. Read-only memory (ROM) is the computer's permanent operating instructions – chips loaded with information permanently etched in during manufacture. Random-access memory (RAM) stores temporarily the new data and programs needed for any computation.

Monitor

External speaker

Microphone

Tablet and cordless pen

Floppy disk

CD

Mouse

Keyboard

Add-in expansion card

Power transformer

Expansion slots

Central processing unit (CPU)

Read only memory (ROM)

Random access memory (RAM)

All information is processed by CPU

Control chips

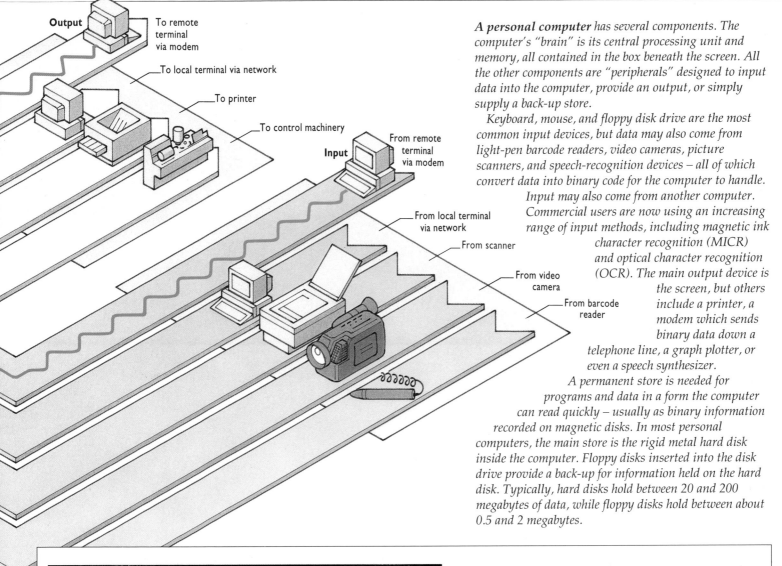

To remote terminal via modem

To local terminal via network

To printer

To control machinery

Input

From remote terminal via modem

From local terminal via network

From scanner

From video camera

From barcode reader

A personal computer has several components. The computer's "brain" is its central processing unit and memory, all contained in the box beneath the screen. All the other components are "peripherals" designed to input data into the computer, provide an output, or simply supply a back-up store.

Keyboard, mouse, and floppy disk drive are the most common input devices, but data may also come from light-pen barcode readers, video cameras, picture scanners, and speech-recognition devices – all of which convert data into binary code for the computer to handle.

Input may also come from another computer. Commercial users are now using an increasing range of input methods, including magnetic ink character recognition (MICR) and optical character recognition (OCR). The main output device is the screen, but others include a printer, a modem which sends binary data down a telephone line, a graph plotter, or even a speech synthesizer.

A permanent store is needed for programs and data in a form the computer can read quickly – usually as binary information recorded on magnetic disks. In most personal computers, the main store is the rigid metal hard disk inside the computer. Floppy disks inserted into the disk drive provide a back-up for information held on the hard disk. Typically, hard disks hold between 20 and 200 megabytes of data, while floppy disks hold between about 0.5 and 2 megabytes.

PICTURES ON COMPUTER

Today many films, such as *Lawnmower Man* (**left**), contain computer-generated sequences. To be processed by a computer, a picture is broken down into tiny pieces called pixels (picture cells). A group of numbers is then assigned to each pixel. Some numbers describe the pixel's position, others its brightness and its color in percentages of red, green, and blue.

Changing the numbers changes the picture, so any bit of the picture can easily be made into a different color, or one element moved to another place. And if a computer is provided with the first and last images from a short sequence, it can quickly fill in the images between.

The rise of robots

Soon robots will spread from factories and warehouses to our shops and offices or even our homes.

Robots in science fiction stories often look like people, with two eyes, two legs, and two arms. But in the real world a robot is just a machine that can learn or be programmed to work, so today's robots look a lot different from humans.

For example, simple industrial robots are just single arms, rooted to one spot, endlessly and mindlessly repeating a series of tasks. They are now a big part of the workforce in all industrial countries, performing jobs once the domain of skilled workers. To train a robot, a human worker merely leads the robot through the actions involved in, say, painting or welding. The human worker performs the task with a special tool linked to a computer, which stores the information about the worker's movements. The information can then be repeated, but this time as actions by robot arms rather than flesh-and-blood arms.

Simple robots have no senses, but more sophisticated robots are being developed that are equipped with TV cameras backed up by clever computer programs to provide vision. They often also have sensors to signal the pressure with which they are gripping an object. These abilities are needed to perform delicate tasks such as assembling parts from components that are mixed up in a box.

If such robots are made mobile, they can move around a factory, mixing safely with the workers. Already AGVs, or

*A climbing robot scales the shiny glass frontage of a modern high-rise building (**left**). It has pneumatic suckers that grip when air is drawn out of them and release when it is blown into them. The air flow and the movements of the legs of the machine are computer-controlled.*

Machines like these could be invaluable in performing dangerous tasks on building sites. Alternatively, they might be used for hazardous operations such as rescue work and firefighting – where they could carry a hose right to the site of the fire.

*A team of robots tirelessly toils in an assembly plant (**above**). The robot's jointed limbs, guided by computers, are able to reach into nooks and crannies. Machines like these can work continuously with breaks only for maintenance.*

This simple robot is nevertheless extremely versatile, capable of being trained to do an enormous number of different jobs. But it is "blind": when working on a task, it has to be presented with identical pieces of work that are in exactly the right position, or it cannot do its job.

automatic guided vehicles, transport goods around some factories and warehouses, following permanent tracks in the floor, obeying commands radioed from a central computer.

Robots will soon mix in wider human society. They may first appear as automatic security guards and firefighters or as rescuers sent into burning buildings. To be useful on construction sites or as cleaners, they will need to be able to negotiate curbs or stairs, as well as to avoid people. So robots with legs may make an appearance. But human–robot hybrids may prove more important, at least in the short run. In the near future, the human body's strength will be amplified by robot limbs: soldiers will be able to vault across battlefields, and people with leg or foot disabilities will be able to walk.

Designed to deal with work inside the highly radioactive zones within a nuclear power station, this huge, fully mobile robot (**above**) can perform tasks in an environment that would be hostile to human life.

Only in recent years have robots become a practical reality. But stories about mechanical people or animals have existed for 2,000 years, although the term "robot" was only coined in the 20th century by the Czech playwright Karel Čapek, who used it to describe the mechanical workers in his 1921 play RUR (Rossum's Universal Robots). Today, robots are popular fictional characters which crop up in books, films, and TV shows.

Modeling reality

The real world is a complicated place and sometimes hard to predict. Coming to the aid of scientists is the computer, which can picture reality's rapid changes.

To make a successful working model of something such as a steam engine, the solar system, or the human skeleton, you have to understand how that thing is put together and how it works. Computers have the great advantage of allowing scientists and engineers to try out electronic "working" models of all aspects of the world. In computer models numbers represent things, and equations represent the relationships between them. The power of the computer to make an immense number of calculations every second allows complex processes to be represented and manipulated quickly.

The range of things modeled is now huge, including patterns of industry and trade, the collisions of subatomic particles, the world's climate, and the birth of galaxies. Scientists can test their understanding by running their computer models, then comparing the results with reality.

When a model is well established and seems to be matching the behavior of a real-life process, its creators can run experiments that they could never perform in the real world. The effects of small changes in the design of a plane, ship, or car can be tested immediately and cheaply. The impact of different levels of pollution on an estuary's ecosystem over a period of years can be compared. The showers of particles to be expected from a subatomic collision, according to different theories of matter, can be predicted, to be compared one day with the results of experiment.

A model that is closer to the everyday meaning of the term is the computer representation of the layout of a building or a town. In these all aspects of a building or a city development can be explored before the first brick is laid. And tourists can visit an ancient city reconstructed in a computer by archaeologists using the remains of the city as a guide.

From modeling the real world, the next step is to enter a computer-created reality and interact with it. Users of virtual reality systems can look around and up and down, and be shown the corresponding view of the computer world. They can be presented with the appropriate sensations on reaching out and grasping or lifting objects. Already the drug designer can manipulate graphics of molecules like a sculptor does clay. And fighter pilots will be equipped with helmets that give them the illusion of seeing all around the sky with radar eyes.

CURVES OF INFINITE COMPLEXITY

Natural textures are often "self-similar," meaning they resemble each other closely on all scales. For example, bare, rocky terrain looks much the same whether it is photographed from a satellite, from a low-flying aircraft, by a person standing on it, or through a microscope. Looking deeper into nature, there is complexity at every level, and the pattern or shape of the natural phenomena repeats itself to look the same at every level.

The newly discovered mathematics of fractals can match nature's complexity with equations that generate infinitely complex curves which look the same however much they are magnified. When these complex fractal curves are used to help create computer graphics, the textures of nature are realistically mimicked (**right**).

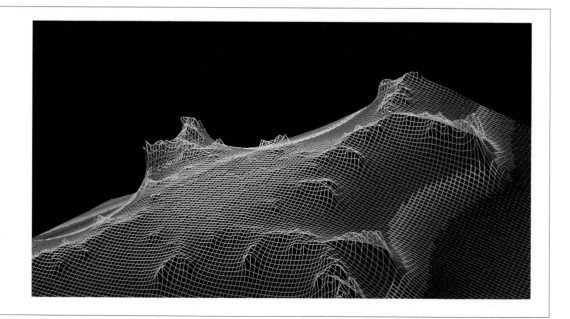

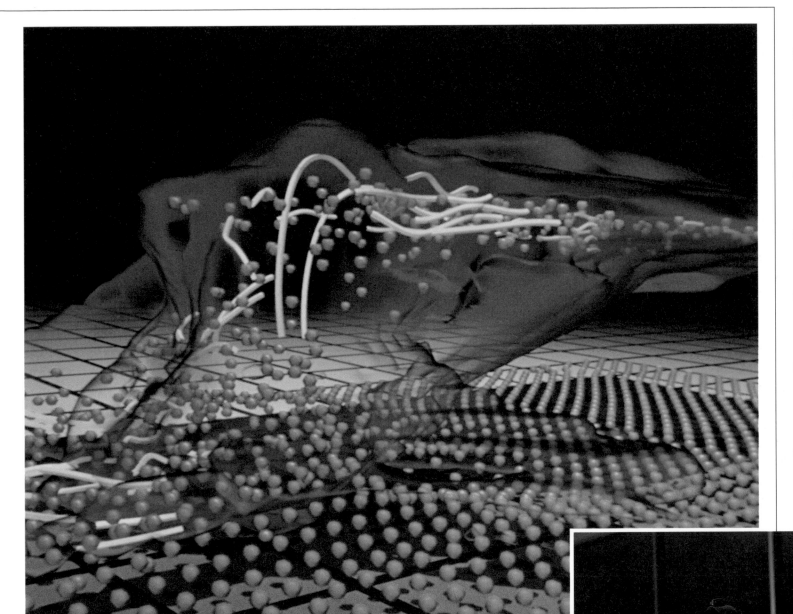

A storm cloud towers into the atmosphere of a computer-simulated world (**above**). The simulation was run on a Cray 2 supercomputer at the University of Illinois. The colored balls show movement of the air: they were all initially in a horizontal plane 1 mile (1.6 km) above the ground, but some have been displaced. Rising air is represented by the orange balls, descending air by the blue ones. The yellow trails show the movement of air during the previous 500 seconds. The computer simulation of the air movement inside a cloud accurately mimics the real life events in a natural storm.

Wearing a virtual reality helmet and glove, the user sees and feels in a world created solely in a computer (**right**). A slightly different view of the scene is presented to each eye, to give a 3D view. Movements of the head are detected by sensors in the helmet, and the computer adjusts the view accordingly. Movements of the glove control an image of a hand in the virtual world.

Mimicking the mind

In the quest for a truly intelligent computer, computer designers are trying to mimic the workings of the human mind.

A chess-playing computer available from most toystores can easily outplay all but the best human chess players. Yet even a child can pick up scattered chess pieces and set them up on the board – a feat far beyond even the most advanced industrial robots. The computer succeeds at chess by churning relentlessly and rapidly through all the possible positions. Humans cannot do this, but they can spot patterns easily, produce the occasional flash of brilliance, and direct the sophisticated movement of body muscles.

This is basically because human brains and computers work in very different ways. A computer has just a few million components wired together in a simple pattern. It can generate tens of millions of signals a second, each traveling at the speed of light, but plods through every calculation step by step. The human brain, by contrast, has 100 billion neurons (nerve cells). Each neuron generates barely 200 signals a second, crawling at only 330 feet (100 m) per second. Yet because each neuron is cross-connected to hundreds of others, and new links are constantly being made, the brain can perform hugely more complex tasks than a computer.

Can computers be radically improved by mimicking the

An organ-playing robot "reads" sheet music and plays the tune on the electronic keyboard. This is everybody's idea of an anthropoid robot, complete with head, arms and fingers, and legs. Professor Ichiro Kato had a serious purpose in building the machine, WABOT-2, at Waseda University, Japan.

Interpreting a TV image and precisely controlling limbs are techniques that must be mastered before robots become "smart," mobile, and fit to leave specialized environments. But human beings remain far superior in these skills.

Computer designers of the 1990s believe that one day they will be able to build a computer or robot that can play, and even compose, music as well as many humans. But there will remain a fundamental flaw in the robot musician; no one can yet imagine a robot that will enjoy listening to the music it plays so accurately.

A neural network computer can learn to identify faces with only 20 seconds of training. It may eventually learn to pick up chess pieces and put them on a board. But will it ever have any fun?

Serial processing   Parallel processing

Input

Output

Output

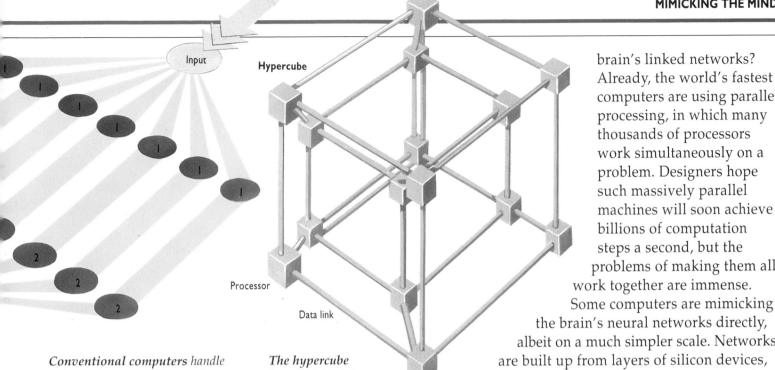

Input

Hypercube

Processor

Data link

brain's linked networks? Already, the world's fastest computers are using parallel processing, in which many thousands of processors work simultaneously on a problem. Designers hope such massively parallel machines will soon achieve billions of computation steps a second, but the problems of making them all work together are immense.

Some computers are mimicking the brain's neural networks directly, albeit on a much simpler scale. Networks are built up from layers of silicon devices, a little like conventional chips, but linked together both horizontally and vertically.

A neural net like this may eventually be able to learn from its mistakes just like (some) humans, continually adjusting the "weighting" of vertical signals until it gets the right answer. In this way, by trial and error, a computer may learn to do anything from deciding on a bank customer's creditworthiness to reading a handwritten letter. It may even be able to tell whether someone is happy or sad just from their video image.

Conventional computers handle tasks one by one in series, so that every tiny calculation must be complete before another can begin. This inevitably places limits on a computer's speed of operation.

But by dividing up the tasks and sending each part through separate processors they can be processed simultaneously. Known as parallel processing, this setup will hugely increase the speed of problem solving.

Some computers have already achieved a type of parallel processing known as "pipelining," where many processors are connected in series, but each works on a separate task.

The hypercube is a sophisticated form of parallel processing. Sixteen individual processors are connected in a network so that each is linked to four neighbors. Messages pass between the processors, and all have access to the main memory.

In any network of processors, the computation is slowed down by the time taken for messages to travel from one processor to another. This delay is minimized in the hypercube layout, in which no message has to cross more than four links to reach any other processor.

NEURAL NETWORKS

Unlike circuits in a computer, every neuron in the human brain is linked to scores of others. Computer neural networks try to mimic this with three layers of silicon "cells," each linked to all its neighbors.
The input layer is like the human retina.
The "hidden" middle layer is like the neurons in the human visual system that process signals from the retina.
The third "output" layer is like cells in the vision areas of the brain that register features such as colors.

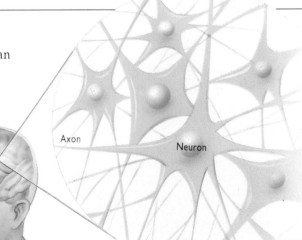

Axon

Neuron

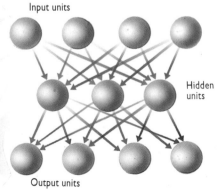

Input units

Hidden units

Output units

Artificial intelligence

The abilities we most take for granted – talking, thinking, and understanding – are the hardest for computers to copy.

Intelligent computer systems are described as those that show flexible, adaptive behavior that no one would hesitate to describe as intelligent if performed by a human being. An area where artificial intelligence has had success is in speech recognition. Simple systems that can recognize spoken commands from a limited set are in use today.

But going further in the direction of understanding language requires vast quantities of knowledge about the human and physical worlds. It gets easier if the field of knowledge is restricted. A computer can converse almost intelligently if the talk is limited to, say, restaurants. But it needs to be equipped with a large set of rules about what happens in restaurants: for example, that customers order food and waiters provide it, that customers tip waiters, but waiters do not tip customers, and so on.

Since 1984, the Cyc project in the USA has been compiling a gigantic database of such facts about the social and physical universe, intended to be a resource for intelligent systems. In the first seven years, nearly one and a half million facts were recorded. By about 1995, Cyc will have stored some 10 million facts; its begetters hope this will prove to be a "critical mass," after which the system will be able to harvest ever more knowledge for itself by browsing through electronic encyclopedias and journals, with little human intervention.

__Genghis the robot ant__ "lives" at the Massachusetts Institute of Technology where it helps with research into artificial intelligence. Genghis is about 1 foot (30 cm) long and each leg is powered by two microchip-controlled motors. The machine can be programmed with the ability to learn to walk. It waves all legs randomly, but within a minute or so learns the combination of movements that best gets it around – a sequence of leg movements used by real insects.

1 Child: The goat has walked onto the bridge. "Trip-trap, trip-trap" sounds like the noise the hooves would make.

Computer: I've never seen "trip-trap, trip-trap" before. If "Trip-trap, trip-trap" is a place, then the goat's hooves have gone there. If "Trip-trap, trip-trap" is not a place, then the goat's hooves have gone onto the bridge. No data about goat's location.

2 Child: The troll has popped his head out from where he was hiding.

Computer: *Either* someone called Out has popped the troll's head *or* the troll's head has popped out.

3 Child: The ugly troll frightens young billy-goat Gruff.

Computer: *Either* the young billy-goat Gruff is so ugly that he and someone called fright nearly fell down *or* someone unidentified is so ugly that the young billy-goat and someone called fright nearly fell down.

The young billy-goat Gruff said that he would be the first to try to cross the river. His hooves went *trip-trap, trip-trap*[1] on the wooden bridge.

Out *popped*[2] the troll's head. He was so ugly that the young billy-goat Gruff nearly *fell down with fright*[3].

"Who's that trip-trapping over my bridge?" roared the troll. A *frightened little voice*[4] replied, "I've heard there's sweet green grass over the bridge."

4 Child: The young billy-goat Gruff replied in a frightened little voice. He had heard that there was sweet green grass on the other side of the river.

Computer: A voice replied to the troll's question. Something had frightened the voice. The voice said that it had heard that there was sweet green grass either on, above, or beyond the bridge.

KNOW WHAT I MEAN?

Almost without thinking, a child can understand a simple story, but a computer has real trouble. This is because comprehension involves having access to a vast amount of knowledge and being able to put that knowledge into a meaningful context.

Using a few lines from a simple story (**left**), a child's understanding is compared with the likely mistakes a typical computer language recognition system might make. Unless the computer is programmed to know all the interrelationships between all the different items of information in a story, it comes up with absurd ideas.

BUGS IN THE SYSTEM

Humans, unlike computers, do not have any difficulty in understanding that each of the pictures (**right**) shows a rabbit. Recognizing a rabbit involves picking out its key features – perhaps the shape of the ears. A cartoon character such as Bugs Bunny shows a rabbit stripped down to its essentials such as ears, teeth, and tail, while a photograph is made up of continuous gradations of color and brightness. But from the mass of information on the photo, a human has no problem picking out the features of the rabbit.

Beyond the chip

Computer power may find its way into all facets of our lives, from running a bath to cooking a meal and washing the dishes.

Dramatic increases in computer power may mean that in the future computers could begin to take over all kinds of tasks once performed only by humans. Inexpensive "low-IQ" computers may well become as normal a part of the average household as light switches. Light switches themselves may become redundant once lights can switch themselves on when someone enters a room and off when the last person has left.

More advanced technology may produce even more startling possibilities. Advanced vision systems, able to distinguish a cup from a sugar bowl, may be combined with sophisticated manipulation devices to produce machines able to set the dinner table – then clear up and wash the dishes.

Computer power should make huge amounts of information and entertainment accessible. Numerous satellites will provide "data highways," carrying TV, phone channels, and information sources, routed via cable to homes and offices.

The hardware that might provide all this can be foreseen in part. Disk drives will be replaced by faster, more compact microchip memories. Compact disks may go the same way. The silicon chip itself may be replaced by optical chips, as copper telephone wires are being replaced by optical fibers.

Already an electronic switch has been made that relies on the movement of a single atom. The aim is to develop such a device so a computer's entire central processor could be built from atom-sized switches. Memory could be stored as patterns of atoms in complex molecules. Such a machine would be incomparably faster than current ones and would pack vastly more memory.

But in a different direction, there is the bizarre prospect of computers made from living tissue. Nerve-cells can already be grown in small numbers in a network to study their functioning. On a large scale, this could form a "brain in a bucket," which might achieve a complexity impossible for a silicon computer.

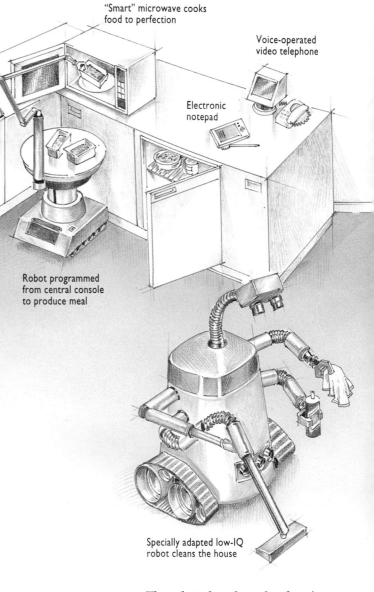

Automated dishwasher loads and unloads itself

"Smart" microwave cooks food to perfection

Voice-operated video telephone

Electronic notepad

Robot programmed from central console to produce meal

Specially adapted low-IQ robot cleans the house

The robot that does the cleaning and helps with all the household chores may well become a reality. Voice recognition systems will enable it to obey commands, while numerous feedback devices will make sure it can respond to changes in its environment.

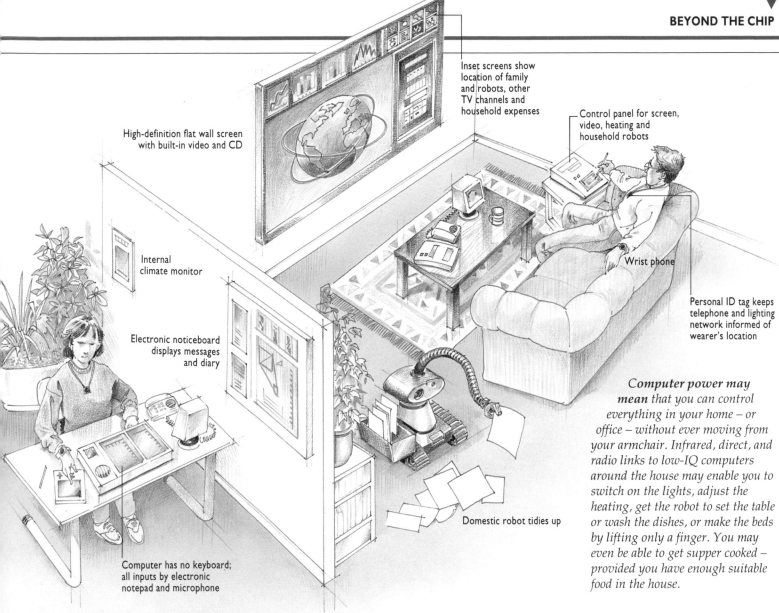

High-definition flat wall screen with built-in video and CD

Inset screens show location of family and robots, other TV channels and household expenses

Control panel for screen, video, heating and household robots

Internal climate monitor

Wrist phone

Electronic noticeboard displays messages and diary

Personal ID tag keeps telephone and lighting network informed of wearer's location

Domestic robot tidies up

Computer has no keyboard; all inputs by electronic notepad and microphone

Computer power may mean that you can control everything in your home – or office – without ever moving from your armchair. Infrared, direct, and radio links to low-IQ computers around the house may enable you to switch on the lights, adjust the heating, get the robot to set the table or wash the dishes, or make the beds by lifting only a finger. You may even be able to get supper cooked – provided you have enough suitable food in the house.

The home of the near future may be transformed as present-day devices are linked and made "smarter" by computer power. Lights may switch on automatically when it gets dark. The heating may come on if it gets cold – or with a phone call from the office.

Simply by asking, you may be able to get the videophone to call a friend or get the music system to put on the music you want at just the right volume. High-definition screens may show electronic newspapers or magazines, and constantly updated video encyclopedias may give you all the information you ever need.

WRITE ON COMPUTER

Intelligent computers are already proving their worth by recognizing handwriting. Clipboard-sized portable computers exist that can take handwritten input. This not only means that a user does not have to learn keyboard skills, but also that the computer can be made use of in situations where speed of input and flexibility of use are essential.

Bibliography

GENERAL

Blackburn, Dr David and Professor Geoffrey Holister (eds.) *Hutchinson Encyclopedia of Modern Technology* Hutchinson, London, 1987

Calder, Nigel and John Newell (eds.) *Future Earth* Christopher Helm, London, 1988

Clark, John O.E. (ed.) *The Human Body* Arch Cape Press, New York, 1989

Hann, Judith *How Science Works* Reader's Digest Association, Inc., Pleasantville & Montreal, 1991

Headlam, Catherine (ed.) *Kingfisher Science Encyclopedia* Kingfisher Books, London, 1991

How is it Done? The Reader's Digest Association Ltd, London and New York, 1990

Macaulay, David and Neil Ardley *The Way Things Work* Dorling Kindersley, London, 1988

The Mitchell Beazley Atlas of the Body and Mind Mitchell Beazley, London, 1976

Sherwood, Dr Martin and Dr Christine Sutton *Hutchinson Encyclopedia of Science in Everyday Life* Hutchinson, London, 1988

Stockley, Corinne, Chris Oxlade and Jane Wertheim *Usborne Illustrated Dictionary of Science* Usborne, London, 1992

THE COSMOS

Galaxies (Voyage Through the Universe series) Time-Life Books, Amsterdam, 1988

Kerrod, Robin (ed.) *The Heavens: Planets, Stars, Galaxies* The Leisure Circle Limited, London, 1984

Moore, Patrick (ed.) *The Astronomy Encyclopaedia* Mitchell Beazley, London, 1987

The Near Planets (Voyage Through the Universe series) Time-Life Books, Amsterdam, 1989

Nicolson, Iain and Patrick Moore *The Universe* Collins, London, 1985

Ridpath, Ian *Longman Illustrated Dictionary of Astronomy & Astronautics* Longman, Harlow, UK, 1987

Ronan, Colin A. *The Natural History of the Universe* Doubleday, London & New York, 1991

Smith, Peter J. (ed.) *Hutchinson Encyclopedia of the Earth* Hutchinson, London, 1986

Strickberger, Monroe W. *Evolution* Jones and Bartlett Publishers, Boston, 1990

Whitfield, Dr Philip and Joyce Pope *Why do the Seasons Change?* Hamish Hamilton, London, 1987

ENERGY

Cooper, Chris and Tony Osman *Colour Library of Science: How Everyday Things Work* Orbis, London, 1984

Dobson, Ken *The Physical World* Thomas Nelson and Sons Ltd, Walton-on-Thames, UK, 1991

England, Nick *Physics Matters* Hodder and Stoughton, London, 1989

Shepherd, Michael *Revise Physics* Charles Letts & Company Limited, London & New York, 1979

Taylor, J.R. and P.M.W. French *How it is Made: Lasers* Faber and Faber, London, 1987

ATOMS AND MATTER

Holman, John *The Material World* Thomas Nelson and Sons Ltd, Walton-on-Thames, UK, 1991

Johnson, Colin *Chemistry for GCSE* Heinemann Educational Books, Oxford, 1987

Lederman, Leon M. and David N. Schramm *From Quarks to the Cosmos: Tools of Discovery* Scientific American Library, New York, 1989

Lewis, Michael and Guy Waller *Thinking Chemistry* Oxford University Press, Oxford, 1986

LIFE

Alberts, Bruce, Dennis Bray, Julian Lewis, Martin Raff, Keith Roberts and James D. Watson *Molecular Biology of the Cell* Garland Publishing, Inc., New York & London, 1983

Attenborough, David *Life on Earth: A Natural History* The Reader's Digest Association Ltd, London, 1980

———*The Atlas of the Living World* Houghton Mifflin Company, Boston, 1989

———*The Trials of Life: A Natural History of Animal Behaviour* The Reader's Digest Association Ltd, London, 1990

Ayensu, Professor Edward S. & Dr Philip Whitfield (eds.) *The Rhythms of Life* Marshall Editions, London, 1982

Burnie, David *How Nature Works* Dorling Kindersley, London & New York, 1991

Downer, John *Supersense: Perception in the Animal World* BBC Books, London, 1988

Friday, Adrian and David S. Ingram (eds.) *The Cambridge Encyclopedia of Life Sciences* Cambridge University Press, Cambridge & New York, 1985

Mackean, D.G. *Introduction to Biology* John Murray Ltd, London, 1977

The Mitchell Beazley Family Encyclopedia of Nature Mitchell Beazley, London, 1992

Moore, Dr Peter, Professor R.J. Berry and Professor A. Hallam *The Encyclopaedia of Animal Ecology and Evolution* Grolier International, Inc., Danbury, Conn., 1986

Whitfield, Dr Philip (ed.) *The Animal Family* Hamlyn, London & New York, 1979

———*The Hunters* Hamlyn, London & New York, 1978

———*Oceans* Viking, New York; Penguin, London 1990

BRAINS AND COMPUTERS

The Mind: Into the Inner World Torstar Books, New York & Toronto, 1985

Blakemore, Colin *The Mind Machine* BBC Books, London, 1991

Blissmer, Robert H. *Introducing Computers: Concepts, Systems, and Applications* John Wiley & Sons, Inc., New York & Chichester, UK, 1992–1993 edition

Fincher, Jack *The Brain: Mystery of Matter and Mind* Torstar Books, New York & Toronto, 1984

Graham, Robert B. *Physiological Psychology,* Wadsworth Publishing Company, Belmont, California, 1990

McLeish, John *Number* Bloomsbury, London, 1991

Marsh, Peter (consultant) *Robots* Salamander Books Ltd, London, 1985

Russell, Peter *The Brain Book* Routledge, London, 1989

Warwick, Roger and Peter L. Williams *Gray's Anatomy* Longman (medical), Edinburgh, (37th ed.) 1989

Whitfield, Dr Philip and Mike Stoddart *Hearing, Taste and Smell: Pathways of Perception* Torstar Books, New York & Toronto, 1985

Index

Acknowledgments

l = left, *r* = right, *t* = top, *c* = center, *b* = bottom

Picture credits
2*tl* Pictor International; 2*tr* Fernando Bueno/The Image Bank; 3*b* N.A.M Bromhall/Oxford Scientific Films; 6*l* George I. Bernard/Oxford Scientific Films; 6*r* James King-Holmes/W. Industries/Science Photo Library; 12/13 Jonathan Scott/Planet Earth Pictures; 18 Gunter Ziesler/Bruce Coleman; 18/19 Gordon Langsbury/Bruce Coleman; 19 R.L. Manuel/Oxford Scientific Films; 20/21 Zefa Picture Library; 23 Stephen Krasemann/NHPA; 24/25 Orion Press/Zefa Picture Library; 25*t* Zefa Picture Library; 25*b* Doug Allan/Science Photo Library; 30/31 Jake Rajs/The Image Bank; 31 NASA/Science Photo Library; 33 Betty Milon/Science Photo Library; 35 Fred Espenak/Science Photo Library; 41 Space Telescope Science Institute/NASA/Science Photo Library; 42/43 NASA GSFC/Science Photo Library; 44/45 Galaxy Picture Library; 46*t* Zefa Picture Library; 46*c* Jeff Spielman/Stockphotos; 46*b* Patrick Doherty/Stockphotos; 47 John Barlow; 48/49 C.Pedrotti/Allsport; 50/51 John Barlow; 52 Patrick Doherty/Stockphotos; 52/53 Yann Arthus Bertrand/Allsport; 54 John Barlow; 55*t* Patti McConville/The Image Bank; 55*b* John Barlow; 56 Steve Allen/Stockphotos; 56/57 Steve Krongard/ The Image Bank; 58/59 Fernando Bueno/The Image Bank; 59 Jeff Spielman/Stockphotos; 60 Zefa Picture Library; 60/61 Thor Larsen/Bruce Coleman; 62/63 Robert Perron; 63 Zefa Picture Library; 64/65 S. Cazenave/Allsport; 65*t* Piers Cavendish/Impact; 65*b* Peter Johnson/NHPA; 66/67 Alex Bartel/Science Photo Library; 67 Black & Decker; 68/69 Electricity Association; 71 Snowdon/Colorific!; 72/73 John Barlow; 74/75 Retna Pictures; 76*l* A-Z Botanical Collection; 76*r* Bernard van Berg/The Image Bank; 77*l* Dick Luria/Science Photo Library; 77*r* Siemens; 79 Earth Satellite Corporation/Science Photo Library; 80 Stephen Derr/The Image Bank; 82 Pamela J. Zilly/The Image Bank; 82/83 Leo Mason/Split Second; 84/85 Tony Mottram/Retna Pictures; 86/87 Laurie Lewis; 88/89 Martyn Colbeck/Oxford Scientific Films; 89*t* Alexander Tsiaras/Science Photo Library; 89*b* Alan Choisnet/The Image Bank; 90 NASA/Science Photo Library; 90/91*t* Heinz Fischer/The Image Bank; 90/91*b* Bernard Asset/Allsport; 91 Kim Westerskov/Oxford Scientific Films; 92/93 NASA/Science Photo Library; 94*t* Doug Allan/Oxford Scientific Films; 94*b* Peter Menzel/ Science Photo Library; 95 John Barlow; 96 The Natural History Museum, London; 97 Philips; 98 Doug Allan/Oxford Scientific Films; 98/99 Zefa Picture Library; 100/101 Pictor International; 102/103 Zefa Picture Library; 103 Comstock; 105 Philippe Plailly/Science Photo Library; 106 A. Caulfield/The Image Bank; 107 Austin J. Brown/Aviation Picture Library; 110/111 Philip M. Derenzis/The Image Bank; 112/113 Burt Glinn/Magnum Photos; 114/115 Patti McConville/ The Image Bank; 115 Clive Corless; 116/117 Peter Carmichael/ Aspect Picture Library; 117 Clive Corless; 118 Allsport; 119 John Barlow; 120/121 Fred Mayer/Magnum Photos; 123*l* Zefa Picture Library; 123*r* Simon Bruty/Allsport; 124/125 W.Kaehler/Zefa Picture Library; 125 John Barlow; 126/127 Richard Packwood/Oxford Scientific Films; 128 John Barlow; 128/129 Tony Stone Assoc.; 130/131 Robert J. Herko/ The Image Bank; 132 Michael Holford; 132/133 Peter Menzel/Science Photo Library; 133 John Barlow; 134 Tony Latham; 134/135 Clive Freeman/The Royal Institution/Science Photo Library; 136/137 Gary Cralle/The Image Bank; 137 John Barlow; 138/139 John R. Ramey/The Image Bank; 143 JET Joint Undertaking; 144/145 Lawrence Berkeley Lab/Science Photo Library; 148*t* Gunter Ziesler/Bruce Coleman; 148*c* J.S. & E.J. Woolmer/Oxford Scientific Films; 148*b* Hans Wolf/The Image Bank; 149 Eric Soder/NHPA; 152 Stephen Dalton/NHPA; 152/153 Dr Jeremy Burgess/Science Photo Library; 153 M. Wurtz/Biozentrum, University of Basel/Science Photo Library; 155 George I. Bernard/Oxford Scientific Films; 156/157 Bruce Ayres/Tony Stone Assoc.; 157 J.S. & E.J. Woolmer/Oxford Scientific Films; 159 Rentmeester/The Image Bank; 160/161 Eric Soder/NHPA; 161 Peter David/ Planet Earth Pictures; 163*t* Leonard Lee Rue III/Bruce Coleman; 163*b* Francois Gohier/Ardea; 164/165 M.P.L. Fogden/Oxford Scientific Films; 165 Kjell Sandved/Oxford Scientific Films; 166/167 John Downer; 167*l* N.A.M. Bromhall/Oxford Scientific Films; 167*r* Georgette Douwma/Planet Earth Pictures; 168 Chris Howes/Planet Earth Pictures; 168/169 Gunter Ziesler/Bruce Coleman; 169 Anthony Bannister/Oxford Scientific Films; 171 Zefa Picture Library; 172 Eric Crichton/Bruce Coleman; 173 Jonathan Craymer/Rex Features; 176/177 Dr Tony Brain/Science Photo Library; 178 Tom Leach/ Oxford Scientific Films; 179 Deni Brown/Oxford Scientific Films; 180/181 Pete Atkinson/Planet Earth Pictures; 181*t* Lex Hes; 181*b* Dr Frieder Sauer/Bruce Coleman; 182/183 Hans Wolf/The Image Bank; 183 Nancy Sefton/Bruce Colman; 184/185 Jim Simmen/ Zefa Picture Library; 185 Adrian Davies/Bruce Coleman; 186*tl* Rex Features; 186*tr* Fiona Pragoff; 186*b* J.L. Charmet/Science Photo Library; 191 Allan Parker/ Planet Earth Pictures; 194 Dr Jeremy Burgess/ Science Photo Library; 196 Bob Martin/Allsport; 198 Valerie Taylor/Ardea; 201 Colorsport; 203*t* Stan Osolinski/Oxford Scientific Films; 203*b* James D. Watt/Planet Earth Pictures; 204 Gerard Champlong/The Image Bank; 204/205 Elyse Lewin/The Image Bank; 205 Camilla Jessel; 208/209*t* Peter Miller/The Image Bank; 208/209*b* Dr Raichie/ Washington University; 210/211 Penny Tweedie/ Panos Pictures; 211 John Barlow; 212 Fiona Pragoff; 213*l* Stephen Wilkes/The Image Bank; 213*r* Barry Lewis/Network; 214 Archiv fur Kunst und Geschichte, Berlin; 216/217 J.L. Charmet/Science Photo Library; 217 Barry Lewis/Network; 219 Steve Dunwell/The Image Bank; 223 The Ronald Grant Archive; 224 Peter Ginter/Network; 224/225 Allen Green/Science Photo Library; 225 Hank Morgan/Science Photo Library; 226 Alfred Pasieka/Science Photo Library; 227*t* University of Illinois; 227*b* James King-Holmes/W. Industries/Science Photo Library; 228 Rex Features; 230/231 Louis Psihoyos/Network; 231*l* The Ronald Grant Archive; 231*r* John Daniels/Ardea; 233 Tangent Associates

Illustration credits
Norman Arlott 202; Eileen Batterberry 177*t*; Richard Bonson 150/151, 154/155, 174/175, 192/193 also brain maps on 194, 196, 198, 201; Bill Donohoe 172/173; Richard Draper 16/17, 42/43, 44/45; Andrew Farmer 4/5, 20/21, 30, 32, 34, 40/41, 55, 57, 58/59, 60, 62, 65, 70/71, 81, 82, 84/85, 87, 106/107, 108/109, 126/127, 146/147, 194/195 also connection icons; Chris Forsey 12/13, 14/15, 24, 33, 72/73, 90, 92, 105*b*, 119, 136/137, 160, 164, 184, 214/215, 220, 232/233; Paul Guest 144/145; Peter Hayman 161*c,b*; Richard Hook 141*bl*; Mark Iley 26*l*, 27*r*, 28*l*, 29*r*, 162, 179, 182/183; Aziz Khan 228/229; Max Kindred 205; Sally Launder 158/159, 176/177*b* 196/197, 198/199, 200/201; Coral Mula 48/49, 51, 56, 128/129, 156, 210; Peter Ruane 97*br*; Mike Saunders 22/23, 26/27*c*, 28/29*c*, 36/37, 38/39; Mainline Design 92/93, 97*tl*, 102/103, 104/105*c*, 110/111, 112/113, 114/115, 116, 122/123, 130/131, 135, 138/139, 140/141, 142/143, 170/171, 218/219; Mark Watkinson 53, 66/67, 68/69, 74/75, 78/79, 89, 98/99, 100/101, 120/121, 124/125, 221, 222/223; Anne Winterbotham 188/189, 190/191, 206/207; Michael Woods 161*t*

Design assistance Roger Pring; Between The Lines; Patrick Nugent; Kate Harkness
Authors David Burnie 16–21, 149–185; Chris Cooper 47–93, 214–233; John Farndon 95–148; Steven Parker 187–213; Robin Scagell 11–15, 22–45